MAIZE AND GRACE

Maize harvest, Tanganyika, 1887.

MAIZE AND GRACE

Africa's Encounter with a
New World Crop, 1500–2000

JAMES C. McCANN

HARVARD UNIVERSITY PRESS

CAMBRIDGE, MASSACHUSETTS, AND LONDON, ENGLAND

2005

Library of Congress Cataloging-in-Publication Data

McCann, James, 1950–
 Maize and grace : Africa's encounter with a new world crop,
 1500–2000 / James C. McCann.
  p. cm.
 Includes bibliographical references (p.).
 ISBN 0-674-01718-8
 1. Corn—Africa—History. I. Title.

 SB191.M2M14 2005
 633.1'5'096—dc22   2004054325

For Harold G. Marcus (1936–2003)
Friend, mentor, and historian of great lives

Contents

Preface

The maize plant *(Zea mays)* seems always to have been within range of my peripheral vision. When I was growing up in the American Midwest, cornfields stood behind the house and surrounded my schoolyards; sweet corn was a staple of my mother's garden and roadside stands, as integral a part of the scene as family and friends. Yet I rarely thought of cornfields as anything more than scenery. Those familiar landscapes have now long been covered by asphalt or shopping malls or the lawns of suburban tract houses. Yet corn—which I here call maize—has moved onto the global stage, especially for Africa. It is not that Africa leads the world in maize production—industrial economies like the United States and China hold that distinction—but Africa, more than any other continent, is dependent on maize as a food source. How maize achieved its current dominance in Africa's fields, in its markets, and in its myriad social expressions is my subject. Whether maize has also offered a blessing—its "grace"—is the question implicit here.

This book, with its topic of the role of maize in Africa, had its specific genesis in 1987, when I was in eastern Ethiopia interviewing farmers to evaluate the impact of the drought rehabilitation programs run by the British aid organization Oxfam. As I tried to reconstruct in interviews with farmers what crops had grown in their fields in the previous five years, I became aware of the rapidity with which maize had become the crop of choice. At least a part of my reaction to the increasing presence of maize was annoyance, and a nostalgia for Ethiopia's older, biologically diverse land-scapes—sadly, and increasingly, a landscape of memory more than of reality. Though I think the farmers shared my longing for that bygone world, the larger question was why these farmers were rapidly abandoning older crops for this New World plant that had always played a rather small role in Ethiopian diet and agriculture before the late twentieth century.

If the visual evidence and farmers' personal stories made the new role of maize obvious, the rationale was less clear. It was particularly perplexing when I considered that in a drought-prone area like eastern Ethiopia, farmers were favoring a plant that was more susceptible to drought stress than any other crop they could have chosen. Moreover, not only were prices for harvested maize considerably lower than those for other possible choices (such as wheat or their traditional teff or sorghum), but maize did not form a major part of their traditional diet. It was a puzzlement.

What eventually became clear to me, there and in other areas of Ethiopia, was that the farmers' deeper rationale was a response to the unstable political situation in which they were living. For them maize was a quick-maturing crop that they could eat after only a few months, at the green, milky stage. And it required much less of their labor—took only one pass of the plow, as opposed to three or four arduous plowings for other crops. After all, these farmers in the turbulent Ethiopia of the mid-1980s never knew when their spouses, sons, or daughters might be taken away from their farms by cadres of the socialist government to work on a road, construct a terrace, or, in the case of a son, forced to serve in the growing national army or some local resistance movement. Or the farmers might have the chance to join a food-for-work program of Oxfam or CARE that promised them flour or oil in exchange for a few days of labor away from their farm. In those days no one knew what tomorrow might bring. Planting maize was thus a quite rational and politic solution to a world turned upside down. Maize was a crop that promised survival at a time when plans were of necessity short-term, and solutions expedient.

A couple of years and research projects later I discovered quite a different take on the meaning of maize. Two influential agricultural economists, Carl Eicher and Derek Byerlee, had edited a book entitled *Africa's Emerging Maize Revolution,* which presented maize's success in southern Africa as a harbinger of an African Green Revolution to match that of South and Southeast Asia a generation earlier. In the cases outlined in their book, maize in its hybrid form seemed a triumph of the science of development, in that the crop yielded much larger harvests than did any other available grains, including the local maize types or the crops endemic to places like Ethiopia. In Eicher and Byerlee's view, southern Africa, especially Zimbabwe, had proved the efficacy of maize as the green engine of economic development. By the mid-1980s, when Ethiopia's farmers were broadcast-sowing local maize seeds in a sign of desperation, 100 percent of the maize sown in Zimbabwe was hybrid seed planted in fertilized rows. Zimbabwe (known until 1980 as Southern Rhodesia) had long had white-owned commercial farms, but in the first flush of independence, black farmers on the Communal Lands rushed to adopt the new vision of maize-based prosperity. In Zambia as well, from the early 1960s to the late 1980s farmers increased their maize production by 400 percent, a massive transformation. Their story too is part of this maize narrative. Were these cases models for Africa's future economic development and food security? Eicher and Byerlee seem to think so. For them maize on a vast national scale was a blessing, a source of grace that could remedy the famine and underdevelopment that were part of Africa's tragic recent history.

Africa's shift to maize also reflects a broader global trend in demand. The Centro Internacional de Mejoridad de Maíz y Trigo (CIMMYT, or International Maize and Wheat Improvement Center), the epicenter for international maize research located outside

Mexico City, estimates that by 2020 the demand for maize in developing countries will surpass that for both wheat and rice. Relative to its 1995 level, annual maize demand in sub-Saharan Africa is expected to double by 2020, to 52 million tons (as cited in P. L. Pingali, ed., *CIMMYT 1999–2000 World Maize Facts and Trends*). Whereas much of the world requires maize as a livestock and chicken feed to fatten meat and poultry for burgeoning urban populations whose income is on the rise, Africa's demand is primarily for maize as food for humans. This peculiar situation of Africa by comparison with other world areas sets the continent's romance with maize apart. Does it offer the grace of a Green Revolution? What does the history of maize in Africa tell us?

I recall as a graduate student in the early 1980s coming across a 1966 book, *Maize in Tropical Africa*, by a University of Wisconsin economist named Marvin Miracle. A peculiar book, I thought, in that its topic was a single crop—and a seemingly mundane one at that—a New World crop in the Old World; not even a grain, but a vegetable that people had come to think of and treat as a grain. Miracle's book was a contradiction—in a way prophetic, but also a product of its time. The book was written in the early 1960s, when newly independent African economies were growing and the world had great expectations for these bold new nations. Maize seemed a measure of those prospects. Miracle's data showed the growing role of maize in many of the most promising young states: Nigeria, Ghana, Uganda, Kenya, Tanzania, Ivory Coast, and Zambia. But the book ended its analysis with the mid-1960s, well before Africa's economic reality ran into the immovable obstacles of the 1970s . . . or later.

A second book also inspired me a few years afterward, this one also about a single commodity. It was Sidney Mintz's *Sweetness and Power,* a book about the substance, history, and meaning of sugar. Mintz is not an economist, but an anthropologist who regards sugar less as a measurable commodity than as a feature of life as seen from the doorway of a Puerto Rican farmhouse. From there he led his readers to see sugar as a food, a crop, a global commodity, a cultural symbol, and a window onto social change. More recently Judith Carney's award-winning book *Black Rice: The African Origins of Rice Cultivation in the Americas* has revealed the role of that crop in laying the cultural and economic foundations for settlement of the New World.

Maize and Grace is in some ways an updating of Miracle's work, though it makes its appearance in a different world and acknowledges a different Africa. Maize is now Africa's most important cereal crop, at once sustaining subsistence and, in other places, providing the seeds of a burgeoning agro-industry, the signpost of modernity. It is at once a triumph of the science of development and the tripwire for its potential failure. Similar to Mintz's sugar, maize is a single commodity, but at the same time it represents breadth of vision. Maize is also a metonym for the environment, in which agriculture is a central feature of the interaction of humankind and nature, as well as being part and parcel of the dynamic landscapes of African agroecology.

This book is broadly cast geographically and thematically; it offers an overview but also looks at events and processes from the farm, fields, or roasting fire. Many of the statistical data cited here identify sub-Saharan Africa as the unit of measure—that is, the continent excluding North Africa and Egypt. None of my case examples come from North Africa, but the Nile Valley, including Egypt, is integral to maize's percolation into the continent and is part of the story. I ask readers to forgive the inevitable oversights.

The goal here is to outline and illustrate maize's historical encounter with the landscapes of Africa over half a millennium—that is, from its introduction in around 1500 to its current apotheosis as Africa's dominant food crop. The implicit question throughout is whether that encounter has been a story of grace bestowed on the Old World by the New, or whether this is a more fundamental human tale of struggle for both sustenance and meaning.

Africa and the World Ecology of Maize

1

Maize expresses its own history not only in its genetic makeup but in its varieties, its agronomic imperatives, its qualities as food, and its peculiar symbiosis with its human hosts and the land they inhabit. Maize is a versatile player, which both shapes and takes the shape of the societies that cultivate it. If we know the plant's genetic endowments and agronomic idiosyncrasies, we can detect in the otherwise disappointingly bare formal historical record some curious details, and even evidence of dramatic historical shifts. Maize comes in five phenotypes—sweet, pop, floury, dent, and flint—yet all its forms derive from a single ancestor domesticated in central Mexico around seven thousand years ago. Though the exact date and circumstances of *Zea mays*'s first cultivation remain a mystery, by A.D. 1500 the Aztec and Mayan civilizations had long called the descendants of that plant maize, meaning literally "that which sustains life," and claimed that the crop was flesh and blood itself. In the modern economies of the United States, East Asia, and Europe it is the ultimate "legible" industrial raw material: industry uses its starches and cellulose in fuel, fodder, paint, plastic, and penicillin.[1] Africa's maize, however, has a different function altogether. Africa is distinctive among world regions in that 95 percent of its maize is consumed by humans rather than being used as livestock feed or industrial raw material.[2]

Why is maize different from other crops? Unlike wheat, rice, barley, and most other cereal crops, which are self-pollinating, maize is an open-pollinating plant. In wheat, for example, the pollen that fertilizes an ovary almost always comes from the stamen of the same plant and has traveled only a few microns to reach its female partner. A maize plant's stamen and ovaries, by contrast, are separated by a meter or more, and the plant produces massive amounts of pollen—about fifty thousand pollen grains per plant—to guarantee its propagation. Moreover, because wheat and rice plants fertilize themselves, the genetic identity of each succeeding generation is virtually identical to that of the first. Fields of wheat and rice, therefore, are for all practical purposes genetically homogeneous. With maize, however, unless humans carefully control the pollination process, the plants exchange genetic materials quite promiscuously with neighboring plants and fields. Thus, all the maize in a given field will differ from the previous generation, a trait that lends the plant a capricious and unpredictable character. Over a few millennia, humans have shamelessly sought to control the licentious tendencies of maize, in order to bend it to their will.

Metaphors aside, the sexuality of maize is, in fact, a dynamic and wonderful force. Because the plant is pollinated by wind rather than by insects, maize produces prodigious numbers of large, thin-walled pollen grains, which often travel tens of meters to adjacent plants or fields but which also may fall onto their own silk and thus fertilize their own ovaries (self-fertilize).[3] When maize plants self-fertilize, the next generation often carries dubious traits such as low yield, susceptibility to disease, and stunted plant size. Yet when cross-pollination takes place, whether by accident or by plan, some among the new generation have quite desirable traits, including vastly enhanced yield and resistance to particular insect pests or diseases. Some plants in the new generation may emerge as very tall specimens or as dwarfs. Such preferred traits often emerge when maize breeders deliberately allow two lines to self-pollinate, and then cross-pollinate them in the next generation to create a hybrid.

Where did domesticated maize plants come from? Geneticists and agronomists have searched for decades to find a wild ancestor for maize in Central or South America with, arguably, little success. Plant geneticists have focused attention primarily on the Mexican plant *teosinte*, perhaps a cousin of maize but probably not its progenitor.[4] Maize, it seems, has always been tended by human hands, its life cycle subject to human manipulation, whether by Meso-American women, Cape Verdean mestizos, or white Rhodesian professional plant breeders. The resulting plant biology offers distinctive insights into maize's agronomic personality and historical workings. Like house mice, English sparrows, and *Anopheles gambiae* mosquitoes, maize requires the presence of humans to survive. Unlike such self-pollinating grains as wheat and rice, maize cross-pollinates and then depends on humans to collect and sow its seeds.

Within its twenty chromosomes, each maize kernel contains part of the genetic record of its long and complex history of human-induced husbandry.[5] Through conscious seed selection, both farmers and professional maize breeders have cajoled the maize plant into altering the time it needs to mature, adjusting its height, increasing its yield, transforming its hard starch to soft (or vice versa), changing the color of its grains, or increasing the percentage of its kernels that will pop in a microwave oven. Each of these features can reside within the genetic makeup of a single land race (subvariety) of maize, thus making each kernel the sum of generations of human selection in corn's American homeland—a range of microecologies that include the Caribbean, Brazilian coastal lowlands, and the Andean highlands. Each of these regions contributed a different set of traits that ultimately made certain maize land races adaptable, once they had made the transatlantic journey to particular ecologies of Africa and the Old World as a whole.

Nestled in new transatlantic ecological crèches after 1500, maize possessed an exquisitely complex set of genetic possibilities that allowed Old World farmers almost immediately to tease out new expressions from its genetic palette, according to local needs, ecologies, and tastes. The terms *rustification* and *creolization* describe the process whereby farmers acquire varieties and then select seeds that suit their own needs for and perceptions of taste, color, or maturity traits. The first few hundred years after the arrival of maize in Africa were a process of farming experimentation and brought a new spate of genetic diversification adjusted to microecologies, agrarian systems, and aesthetics. In more recent times quite different ideas, derived from modern agricultural lore, have come along to reverse that diversification.

The key concept in modern maize cultivation is what plant science calls heterosis, or hybrid vigor, which is the phenomenon caused by the interaction of favorable genetic materials as manipulated by professional maize breeders to produce hybrids. Hybrids are the result of crossing, once or more commonly twice, two or more inbred (self-pollinated) genetic lines to produce heterosis, in order to increase yield or produce other desired traits. To maintain those characteristics, however, farmers must use new hybrid seed for each season's planting or risk deterioration in the plants' desirable traits. Depending on the process used, maize breeders call the resulting seed varieties single-cross, double-cross, top cross, or three-way.[6] In 1960, crop scientists from colonial Rhodesia developed SR-52, the world's first commercially viable single-cross hybrid, to benefit a particular ecology and a particular economic group, southern Rhodesian white commercial farmers. Maize breeding thus is often a political as well as an agronomic undertaking.

An alternative, less high-tech version of improved maize types is *composites* or *open-pollinated varieties* (OPVs). These are the result of selection and combination of several desired traits from self-pollinated plantings with the aim of producing a uniform improved crop. While they rarely are as productive as hybrids, OPVs can be replanted by farmers using seed from their own fields. A common metric for assessing agricultural change on African farms in the late twentieth century was the proportion of improved maize (hybrids and OPVs) used by farmers in a given place. In the 1970s and 1980s farmers in southern Africa adopted hybrid maize on a large scale. Improved varieties reach farmers via national seed enterprises, private seed companies (such as the United States–based Pioneer Hi-Bred Seed Corporation), or government agencies. *Recycling* is the term used when farmers attempt to replant their hybrid seeds in the following year, a practice that results in diminished yields and plant quality. In many parts of Africa the lack of farm credit and cash income has tempted farmers to recycle seeds, against the advice of extension agents and seed companies. It remains to be seen whether African farmers' adoption of hybrid seeds is a sustainable change or a temporary blip.

As a food plant, maize has historically had a split personality, appearing in diets in some times and places as a vegetable crop from the garden and at other times cultivated in the field as a grain. On a farm it can be either, or both, being defined as much as anything by its function. People eat maize, as a household garden crop, at its green, milky stage—boiled as a snack or roasted on the cob. Farmers broadcast corn seeds for a field crop onto prepared plots and then harvest the dried ears; women grind its kernels into flour. In a strict nutritional and physiological sense, however, maize is a vegetable rather than a grain, offering vitamins A, C, and E (these vitamins being one of the ways in which a vegetable is defined nutritionally) but lacking the lower B vitamins that characterize a true grain, such as sorghum or wheat. Corn is high in carbohydrates but low in usable protein, especially the vital amino acids lysine and tryptophan; the leucine in corn blocks the human body's absorption of niacin, a vitamin whose absence causes protein deficiency.[7]

When sown as a field crop, however, maize assumes the mantle of a grain, often replacing true grains like wheat, rice, sorghum, or millet in the fields. In modern commodity markets maize also assumes the status of a grain; commercial farmers cultivate, harvest, process, and store it as though it were a grain. Having by the late twentieth century approached the status of a monocrop on many African farms, maize has now overwhelmingly and permanently taken on the dietary characteristics of a grain. Its flour and its kernels then serve as the raw materials for beer or porridge—the latter variously known by the names *papa* (South Africa), *sadza* (Zimbabwe), *ishimi* (Zambia), *gunfo* (Ethiopia), *kenkey* (Ghana), *ugali* (Kenya, Uganda, Tanzania), and many others. While Africans tend to view this stiff maize porridge as their own distinctive food, it is in fact the same stuff consumed in other cuisines variously as grits, polenta, or mamalinga.

African food ways, however, differ from those in other maize-growing areas of the world. Food items derived from maize in Africa are most often boiled or cooked (Ghana's kenkey, Ethiopia's nifro, or Kenya's ugali), whereas in the Americas they are baked or fried (tortillas, cornbread, and hushpuppies). Moreover, the two main types of food maize, flint and dent, are used for distinct types of food. Dent maize is soft and floury and best used for soups and stiff porridges like papa and sadza. Flint maize, by contrast, contains harder starches in its endosperm and appears in local diets in the form of gruel or a type of couscous that replaces rice. Industrial mills, however, invariably prefer—and insist on—the softer dent type, which grinds uniformly in their steel roller mills and causes less injury to the machinery. By contrast, in Malawi, where the urban market is smaller, women have traditionally preferred the flinty types because they suffer smaller losses in storage and give better results during hand-milling.[8] Most African consumers still prefer the texture and flavor of flint maize flour, when they can get it. To cite a non-African example, Italian polenta is made exclusively from flint maize, which amounts to only 15 percent of Italy's total national production.

As a grain, maize yields more food per unit of land and labor than any other. No other cereal can be used in as many ways as maize. Virtually every part of the plant—including the grain, leaves, stalks, tassels, and even roots—can be put to human use.

Depending on their own views of its uses, farmers will choose their own local versions of desirable traits, colors, and textures. In Ethiopia the especially thick stalks of the wildly popular BH660 hybrid variety make particularly good cooking fuel. In Nigeria the variations in local consumer preferences are renowned, neighboring villages having historically preferred quite different color combinations and flour textures.

Yet to those in Africa and in the nonindustrial world who are seduced by the obvious virtues maize can offer, it has also revealed a darker side. The growing plant is highly sensitive to deprivation of water, sunlight, and nitrogen; maize kernels rot easily in tropical storage. Even a few days of drought at the time of tasseling can drastically reduce the yield at harvest. Maize monocultures are thus extremely vulnerable to environmental shocks, especially drought, but even in the best of times a maize-based diet may impoverish the bodies of those who depend too heavily on it for food, and over the long haul such a diet can result in deficiency diseases such as pellagra (a disease caused by vitamin deficiency) and kwashiorkor (a disease caused by protein deficiency). The end result is that in planting maize, commercial farmers and peasant families alike (especially women—African maize is largely a woman's crop) walk a slender tightrope of risk. Still, in the past half millennium the cultivation of corn has continued to spread across Africa, from rain forest plots to cocoa farms and from remote villages to urban vacant lots.

By the first decade of the twenty-first century a tidal wave of maize had engulfed Africa—or at least all but its driest and wettest crannies—in the process supplanting historical African food grains as sorghum, millet, and rice. The recent spread of maize has been alarmingly rapid, although the historical and social implications of that change have received scant consideration from media, social scientists, and policy makers. In southern Africa maize has become by far the most important staple food, accounting for more than 50 percent of the calories in local diets; in Malawi alone,

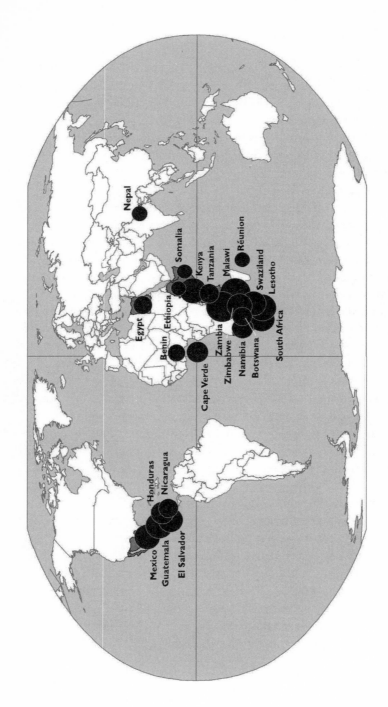

1. Percentage of maize in total national diet, 1994.

maize occupies 90 percent of cultivated land and represents 54 percent of Malawians' total caloric intake. Far from considering it a crop of recent origin, imported from the New World a mere half a millennium ago, Malawians of the late twentieth century stated, "Chimango ndi moyo" ("Maize is our life"), and called their favorite variety maize of the ancestors *(chimango cha makola)*.[9] In the modern landscapes of neighboring Lesotho the impact is even greater: Lesotho's national diet exhibits the world's highest percentage of maize consumption (58 percent of total calories), though Zambia is nearly as dependent on maize as its smaller neighbor. Like Zambia, Malawi, and Lesotho, many of Africa's agrarian and urban landscapes have come to bear the imprint of this single crop.

The consequences of having maize at the leading edge of an African agrarian transformation have been ambiguous: of the twenty-two countries in the world where maize forms the highest percentage of the national diet, sixteen are in Africa.[10] Moreover, the top three countries on this global list are all in Africa (Zambia, Malawi, and Lesotho), surpassing even Guatemala and Mexico, the homelands of maize. In East Africa as a whole, maize accounts for 30 percent of all calories consumed; on the world list, Kenya and Tanzania are sixth and fifteenth in maize consumption, by percentage. In South Africa, maize-growing areas represent 60 percent of all land planted in cereals, and maize 40 percent of all calories consumed. Ethiopia, despite being a world center of crop genetic diversity, now produces more of this New World grain than of any other food crop, including renowned native cereal crops like teff *(Eragrostis teff)* and finger millet *(Corcorana abyssinica)*.[11]

Africa's Green Revolution?

While the Americas produce more of their native crop in total volume, the overall impact of maize globally may be greatest in Africa, where its expansion as a major food source has paralleled the continent's economic and nutritional crises. Since its first arrival with missionaries, merchants, and slave traders, the crop has expanded its domain rapidly. In the past two decades, rapid advance of maize as a major food crop in Africa has caught the imaginations of agricultural economists and international policy planners, who see this

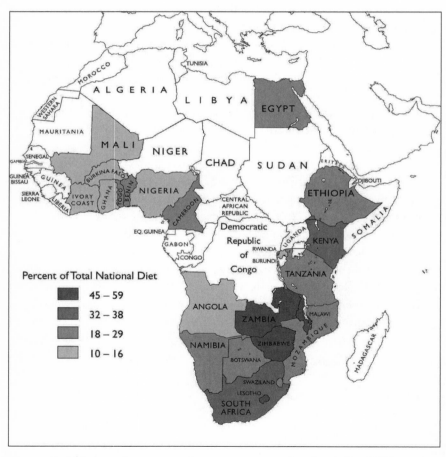

2. African maize consumption as percentage of national diet, 2002.

as an agricultural sea change that could rival Asia's Green Revolution of the 1970s.[12] Entrepreneurial economists see the dominance of maize in Zimbabwe, Kenya, Malawi, and South Africa as a free-market economic miracle. Nobel laureate Norman Borlaug, icon of Asia's wheat-and-rice-based Green Revolution, has also taken up the cause. He argues, through the organization Sasakawa Global 2000, that the technologies and new crop varieties to launch Africa's own Green Revolution, with maize adoption as its most visible expression, are already in existence.[13] These bold claims suggest the need to understand maize's historical role in Africa more fully.

The story of African maize tells us a great deal about the continent's distinctive physical and cultural environments, as well as the many ways Africa has contributed to the global (and globalized) phenomenon of maize. Maize's increasing domination of Africa's agrarian landscapes has wrought a broad array of changes and conundrums. Despite Africa's enormous diversity of cultures, ecologies, and aesthetic sensibilities, in recent years Africans have seemingly reached a consensus that maize is their favorite food and that its color should be white, not yellow, red, or blue.

At the end of the twentieth century the world planted 140 million hectares of maize each year. Of that total, 96 million hectares were in developing countries, a category that includes all of sub-Saharan Africa. Although 68 percent of all land planted with maize is in the developing world, it accounts for only 46 percent of the world's six hundred million tons of production (1999). The discrepancy reflects the fact that the average maize yield in industrialized countries is eight tons per hectare, but in the developing world less than three.[14]

Though maize is a New World plant, Africa in many ways is its ideal setting. Of the seventy million hectares of maize planted in nontemperate and tropical environments, 65 percent are in the tropical lowlands, 26 percent in the subtropics and midaltitude zones, and 9 percent in the tropical highlands. African maize farms occupy 45 percent of the world's subtropical and midaltitude maize-growing area, the plant's favorite habitat. This agroecological zone includes the highly commercialized agriculture of Zimbabwe, South Africa, and Zambia as well as peasant-based production in Ethiopia, Malawi, and Ghana. Sub-Saharan Africa accounts for a fourth of the world's tropical regions planted in maize. As a tropical and subtropical production area, Africa faces the difficulty that the miracle maize varieties that bring high yields in the temperate industrialized world are of little use in tropical and subtropical climates. Moreover, Africa's production is more than 90 percent white maize intended for human consumption, rather than the yellow maize used for livestock feed and industrial purposes that dominates world maize markets. On average, Africa expends more than three-quarters of its maize for human consumption (well over 90 percent if industrial South Africa is excluded). In the high-income countries of North America and Western Europe only 4 percent goes for human consumption.[15]

World maize harvests grew like Topsy in the last half of the twentieth century, in both industrialized countries and the developing economies of Latin America, Africa, and Asia. The numbers are mind-numbing but tell a story of their own about the modern global economy of food. In 1973 more than half the total world area planted in maize (111 million hectares) lay in the developing world, yet those countries accounted for only a fourth of all maize produced. Of this total, Africa accounted for 17.2 percent of the area planted but less than 8 percent of the world's total maize supply.[16] A quarter century later, in 1997, the world produced 580 million tons of maize in toto, almost half of which came from developing countries (259 million tons)—a major change in global agriculture. A good portion of the developing world's increasing share of the world harvest came from improving yields in Africa. In the 1960–1990 period African maize yields improved by only 1 percent annually, but in the 1991–1997 period they improved by 2.9 percent—almost twice as fast as in Asia, but slightly less than in Latin America.[17] Moreover, the rates of increase for both area planted and total maize production in Africa were dramatic.

The geography of Africa's maize also broadened and shifted over the past half century. From 1997 to 1999, eastern and southern African countries produced more than 23 million tons, while central and West African farmers harvested slightly less than half that amount (11 million tons); North Africa (in this case, primarily Morocco and Egypt) produced only 6.4 million. Hidden within these cold statistics are some enigmatic trends that cry out for deeper inquiry. In the century's final decade Ethiopia raised its maize production by 12.3 percent a year, Mozambique by 14.5 percent, and Chad by 17 percent. Zimbabwe, which had raised its production by 6 percent a year in the 1950s, produced almost 1 percent less a year in the decade from 1989 to 1999. Overall, eastern and southern Africa devoted 41 percent of their total cereal-growing area to maize, but West and central Africa only 21 percent. Appendix Table 1.1 merits considerable attention for what it says about changes over time and transformation of national economies. Beneath these numbers, however, lies a still more telling set of stories about environment, politics, farm life, and the culture of farmers themselves.

Though maize arrived in Africa five hundred years ago, it is nevertheless a relative newcomer in a very old and complex environmental setting. Africa is the oldest continent in human terms, but if we go even further back, in geological terms it was the largest fragment of what was the original geological land mass, Gondwanaland. The soils and geomorphology of Africa show both its age and also recent episodes of volcanic upheavals, deposition of sediments, and the action of water and wind on the land's surface. In altitude, the extremes range from 120 meters below sea level in the Danakil depression near the Red Sea to 5,895 meters above sea level at the peak of Mount Kilimanjaro. In northeast Africa, the geological domes that extended from Ethiopia to Tanzania split open 750,000 years ago to create the Great Rift Valley; the Great Lakes formed when the Eastern Rift Dome lifted the earth's crust and left behind a series of long, deep lake basins stretching from south of the equator to (and including) the Red Sea itself. The highlands of northeast Africa form what biologist Jonathan Kingdon calls a fractured dome.[18] While only a small fraction of Africa's land mass lies above 1,500 meters, half of those highlands are in the region of Ethiopia and Eritrea, a concentration that gives the continent as a whole a tilt from northeast to southwest. Forty percent of Africa's land has a slope of more than eight degrees, resulting in the movement and redeposition of soil by rain and river systems—erosion. Erosion is not so much a recent crisis as a consistent and inexorable historical process that defines the areas where maize has found an accommodating ecological niche.

Viewed from a satellite, the continent's major features are vast alluvial plains broken by the branching of old stream beds and living rivers. At ground level the textures and hues of the different soils are more visible. Africa's soils vary dramatically in color, chemistry, and structure, ranging from light, sandy arenosols to heavy, poorly drained black "cotton" soils (vertisols). That something over a quarter of the continent's tropical soils are acidic is significant, since for farmers these soils pose special problems. Not only are acidic soils deficient in phosphorus, calcium, and magnesium, but they often contain toxic levels of aluminum.[19] Even more common in Africa are the red porous laterite soils, which are liable to lose nutrients and are low in nitrogen and phosphorus (both critical to maize plants' yield). These soil conditions are all local phenomena that allow us to make few generalizations. Such variations, though they are a nightmare for modern industrial agriculture, which seeks uniformity and economies of scale, have encouraged African farmers to practice agriculture as a craft rather than an industry.

It is less accurate to say that Africa is losing soil than that the soils move and redeposit themselves. In placing the emphasis on soil loss, standard calculations of erosion rates, which have been grossly exaggerated for places like the Ethiopian highlands, have underestimated the historical effects of soil movement and new soil formation.[20] Over time, farmers have adopted planting strategies and chosen crops to suit Africa's variegated soil landscapes. Historically, African farmers sowing maize have dealt with the variation by creating patchwork plots that reflect soil and crop types—that is, by matching crops to soils and climate conditions. In the long transformation of maize from a household garden crop into predominantly a field crop, the trend has been to insist on changing the soil and agroecology through the use of agrochemicals such as nitrogen fertilizer, herbicides, and pesticides to suit the crop, rather than to conform the crop to the field. This historical evolution from local variation to homogenization has been a defining factor in the history of maize in Africa.

Beyond geomorphology and the genetic diversity in vegetation and animals, the landscapes of Africa and maize's role there reflect changing patterns of climate. Unlike in temperate zones, where growing seasons and life cycles respond most directly to fluctuations in temperature, in Africa the rhythms of life reflect primarily the availability of moisture, especially rainfall.[21] Africa's annual patterns of rainy and dry seasons, humidity, soil moisture, and length of growing season result from the yearly rhythms of global cyclonic winds, ocean temperatures, and the earth's rotation around the sun. In following the tilt of half the earth toward the sun in summer and away in winter, the anticyclonic and trade winds set the yearly cycle of rainy and dry seasons. This shifting zone of rain-bearing turbulence, which climatologists call the Inter-Tropical Convergence Zone (ITCZ), sets a bimodal (two-part) pattern of seasons, one wet and one dry, which characterizes the continent as a whole. This seasonality, however, also produces subtle variations from year to year. In certain places, elevation, topography, and global climate anomalies such as El Niño (or ENSO), La Niña, and tropical Atlantic surface circulation signal drought conditions.[22] The wind circulation that brings moisture to the regions of Africa located north of the equator moves from south to north of the equator as the earth tilts toward the sun in the summer months (June to September). The turbulent movements north of the equator bring summer rains in the northern hemisphere. Africa's annual weather cycle is predictable in broad seasonal terms, even if it is at times erratic from year to year and from locale to locale.

The onset of the rains has a remarkable effect. Within two weeks, brown, lifeless landscapes turn green, seeds germinate, and chemical reactions within soils make nutrients available to plants. From December through March, air masses from the north dominate, creating a long dry season as the rain-producing ITCZ turbulence moves south of the equator. In the dry season, fields ripen for harvest, pasture grasses shift into dormancy, and livestock migrates to pasture near water sources. The dramatic seasonal movement of wildebeest in the Serengeti is a well-documented example of this annual process. The effect on farms is more subtle: maize that was once cultivated as a part of swidden (slash-and-burn cultivation) is now commonly incorporated into permanent fields modified by inorganic fertilizer and pesticides. Seasonal cycles also affect planting times and the choice of maize cultivars to suit the normative length of growing season. But African rainfall always presents unexpected surprises. The interaction of the ITCZ with local topography results in moist slopes or rainfall shadow effects that have repercussions for local crop choices, land values, and landscape texture. The fluctuation of maize yields in southern Africa in the early 1980s shows this effect most dramatically, for the small farms on marginal lands occupied by black farmers have registered the changes first and hardest. Yet even the large commercial farms on better-watered land have eventually felt the pinch of declining yields that (while they bring high prices in the cities) force countries like Zimbabwe, Malawi, and Zambia to import maize.

The great fecundity of the maize plant in good weather conditions is as remarkable as its decline when the rains fail. Drought is the cumulative effect of several years of short or delayed rains, historically a common occurrence in much of Africa. Drought is also a climatic fact of life that greatly influences the geography of maize planting and the ecological niches in which farmers have become

3. Abbay (Blue Nile) Falls, Ethiopia, September 1973 (rainy season).

4. Abbay (Blue Nile) Falls, Ethiopia, May 1974 (dry season).

dependent upon the crop. The quick maturity of some maize types allowed them to "escape" drought—that is, they did not suffer from the rains' early cessation—and to provide food for the household in the hungry season before other crops are available. Less commonly, some types of maize cope with drought by adapting the timing of their pollination (the appearance of tassels) with the appearance of corn silk to receive the pollen.

As a continent, Africa has the world's sharpest seasonal swings from wet to dry. Vegetation patterns, animal movements, and human economies in Africa have adjusted to this cycle over several millennia. In the larger geological time frame, Africa's climate has been more often dry than wet and more often warm than cool. The cereal crops native to Africa, such as many varieties of sorghums and millets, or teff and eleusine (finger millet) in Ethiopia, have adapted to seasonal conditions and periodic drought over a millennium or two. Exotic crops like cassava and bananas—and maize—that originated in either the New World or Southeast Asia, owe their popularity among African farmers to the success with which they have fit into the seasonal systems.[23]

These long-term fluctuations have been significant during epochs of human history in which African social institutions and economic strategies evolved with the experience of human communities and individual farmers, hunters, and pastoralists. Historian John Iliffe generalizes that Africa's agricultural systems were historically mobile, a strategy for adapting to the environment rather than transforming it. The original maize germplasms that came to Africa from Brazil and the Caribbean were largely short-season crops well suited to either moist forest zones or drought-prone savanna areas. By contrast, the modern agricultural preference is to alter (that is, homogenize) local soils and moisture by using inorganic fertilizer, irrigation, pesticides, and so on, to suit improved cultivars, and to manipulate local ecologies to accommodate seeds developed for use across a geographic spectrum. Only recently have large-scale investments made it possible to adapt for local African ecologies new crop varieties that are drought-resistant, say, or tolerant of particular soils.[24]

The Ecology of the Maize Plant

Maize plants exhibit particular sensitivity to African climate patterns. Geographically and historically, rainfall has been the principal factor limiting the types of crops and the times they can be planted in Africa, but temperature imposes additional constraints. Maize grows best at temperatures ranging from 24° to 30° C (72° to 86° F). Temperatures higher than 32° C (90° F) interfere with the plant's physiological processes. Cool periods retard germination, thereby causing a longer maturation period, delayed tasseling and silking (pollination), and late appearance of the green ears that are so welcome during the hungry season before the main cereal harvest. In many areas of the tropical lowlands, therefore, extreme temperatures—rather than moisture—set the limitations on the planting of maize.[25] Ecologies for potential maize production in southern Africa reflect the confluence of certain characteristics of temperature, rainfall, and elevation.

Maize reflects the limitations of Africa's physical environment most directly in the crop's response to moisture stress. Overall, a marginal rain-fed maize environment in the tropics is one where seasonal rainfall is less than 500 millimeters, and in the highlands it is one where rainfall does not exceed 300 to 350 millimeters. In physiological terms, drought stress affects the maize plant's grain yield at three critical stages of its growth: 1) early in the growing season, 2) at anthesis (tasseling), and 3) during grain filling, as the kernels mature on the ear. In regions of early drought stress, farmers have the option of replanting a type that requires less time to reach maturity or a different crop that matures quickly, as was done during the 2002 drought in central Ethiopia. Midseason drought can be especially devastating, since the plant is particularly vulnerable during its period of flowering, when contact between its silk and pollen complete the cross-pollination process. Research shows that drought stress on maize plants a week before anthesis reduces grain yield by two to three times more than it would at other growth stages.[26] In effect, drought stress on maize manifests itself primarily not in lower-than-average annual rainfall or in even seasonal variations, but rather because of insufficient moisture in the four-week period around flowering and silking. Unlike with other cereals, in hermaphroditic maize the male and female parts are separated by considerable distance (about a meter). Drought delays silking, because in underdeveloped leaves photosynthesis is reduced. For sub-Saharan Africa a strong correlation exists between average national maize yields and total rainfall during the growing season, for eastern and South Africa in particular, and somewhat less dramatically for West Africa.

In short, in Africa maize's potential as a food source and cash crop is a function of distinctive patterns of rainfall for that continent. Drought has had quite different effects on different maize farmers and maize economies. In the Horn of Africa, drought and famine have been an omnipresent specter for subsistence farmers, particularly those far from markets and sources of off-farm income. In the mid-1980s, for example, drought in the Horn of Africa combined with radical state interventions in rural economy prompted cultivators to seek low-labor, short-season crops, and a rapid expansion in maize cultivation resulted. By contrast, in the more recent drought of 2002, farmers quickly abandoned maize in favor of traditional grain crops that could tolerate delayed and erratic rains.

Drought's effects in southern Africa were different in kind if not in scale. There, maize cultivation on communal lands and on commercial farms emerged within a cash economy where credit was more available than in the Horn. Drought affected profit margins and overall production, but it did not cause famine, since marketing networks and the cash-based economy provided what Amartya Sen has called cash entitlements.[27] Relative to other crops, paradoxically, maize made advances in both situations. Despite its physiological vulnerability to drought conditions, the political ecology of agriculture in the mid-1980s fostered the expansion of the crop both on Ethiopia's peasant farms and in Zimbabwe's capitalized agriculture. In the economic crisis of 2000 to 2004 in Zimbabwe, inflation, a currency shortage, and drought caused farmers to abandon "modern" varieties and to recycle hybrid seed or to seek less productive but more reliable types of maize seeds.

Maize's Long Journey

Farmers, seed companies, and international agencies (such as CIMMYT) have spread the gospel of maize in its modern form as a grain, in hybrid and open-pollinating varieties that are products of deliberate commercial breeding programs sponsored by governments, seed corporations, and international agencies. Maize is reducible to its genetic elements, a set of building blocks for experimental recombination. Its genetic components can also be multiplied to produce plant populations that occupy huge, uniform cultivations extending to thousands of hectares, a scale that appeals to centralized state planners and corporate farms, and equally to small farmers on half-hectare plots. For better or worse, modern genetic alchemy has transformed maize from an obligingly adaptive vegetable crop to a hegemonic leviathan that dominates regional diets and international grain markets. In contrast to early genetic selections by native Americans—who adapted maize, its colors, and its textures to the vagaries of locality—modern plant breeders choose reliability of character (especially yield) and then seek to manipulate the local ecology with nitrogen, irrigation, and herbicides to achieve uniformity. Modern human management has thus produced a plant that anticipates a predictable ambience and has only a limited ability to conform to diverse local landscapes, soil, and climate.

Maize's metamorphosis from an exotic sixteenth-century visitor to a thoroughly African crop was a long one, entailing a journey from vegetable to grain, from garden to field, from curiosity to staple. The next chapter picks up the historical narrative in greater detail, starting with the sixteenth century, and pursues the crop's shift in scale from local to continental.

Naming the Stranger:
Maize's Journey to Africa

2

After the opening of the Atlantic basin to trade, cultural exchange, and violent exploitation, the Old World was for maize a tabula rasa. Maize arrived in Africa after 1500 as part of the massive global ecological and demographic transformation that historian Alfred Crosby called the Columbian exchange.[1] The great irony, of course, is that the same Atlantic economy that wrenched captives from Africa to supply labor for the preindustrial economies of scale in the New World also provided Africa with new cultigens (cassava, beans, potatoes, and maize) that reinvented Africa's food supply over the next half millennium. There is, however, little documentary evidence of what must have been a conscious process of Europeans' or Africans' first introducing the maize plant to Africa and of African farmers' responding to it. The importation of maize seeds to various parts of Africa went unremarked, though it certainly was not unremarkable.

The first documentary reference to the presence of maize in Africa may be that of an anonymous Portuguese pilot who in 1540 described already well-established cultivation on the Cape Verde Islands: "At the beginning of August they begin to sow grain, which they call Subaru [zaburro], or in the West Indies mehiz [sic]. It is like chick pea, and grows all over these islands and along the West African coast, and is the chief food of the people."[2] On the island of São Tomé, further south, another Portuguese pilot reported in the mid-sixteenth century that the island's slave traders fed their captives on "zaburro, which we call maize in the western islands and which is like chickpeas."[3] Dutch traveler and writer Olfert Dapper remarked in 1668 that maize had earlier been carried from the West Indies to São Tomé and thence to the Gold Coast (where he saw it): "First of all there grows there Turkish wheat, which the Indians call mays and which was first brought from the West Indies where it is plentiful by the Portuguese to the Island of Saint Thomas and which was distributed thence along the Gold Coast for consumption by the blacks."[4]

By the middle of the seventeenth century, European references to maize in settings in West Africa became more commonplace. French scholar Dominique Juhé-Beaulaton has cited in great detail historical *récits de voyage* that describe the appearance of maize on West Africa's Gold Coast beginning in the early seventeenth century. She notes that by the late seventeenth century, maize had largely replaced both millet and sorghum along the West African coast. Presumably, West African farmers had already been exploring the new plant's possibilities, as a food and as a farm crop that filled a distinctive agronomic niche, and had come to prefer the new crop to the indigenous millets and sorghums. By the eighteenth century, maize was the principal *céréale cultivée* in the region. In 1795 Mungo Park found that the people around the Gambia River cultivated it in considerable quantities. Only in two areas of the Gold Coast, in the Volta River delta and at coastal Axim, did rice remain the dominant cereal.[5]

While most twentieth-century writers take these early accounts as evidence that the Portuguese had been the agents of introduction, the Portuguese claim no such thing, or at least they remain mute on the subject.[6] The Nigerian evidence for multiple points of entry, including a minor Portuguese role in the introduction of maize, is much more illuminating, even if more complex. The veteran ethnographer of the Yoruba William Bascom was "exasperated" by his Ife informants' insistence that maize had always been with the Yoruba, and the linguistic gadfly M. W. D. Jeffreys took early Portuguese references to *milho zaburro* (which could easily be translated to mean any grain) to signify that maize had been in Africa since pre-Columbian times.[7] A team of British historical linguists, including the indefatigable Roger Blench, has offered more precise evidence from language and pointed out that there is no direct evidence that the Portuguese played a role in introducing maize to what is now Nigeria. The linguistic patterns they cite, in fact, seem to suggest that maize arrived primarily during long-distance trade with the north, via Bornu (in present-day Chad) and lesser routes from the west and northwest. In only one case is there linguistic evidence of a possible Portuguese role, and that is the case of Isekiri in southwestern Nigeria. Pieter de Marees's 1602 en-

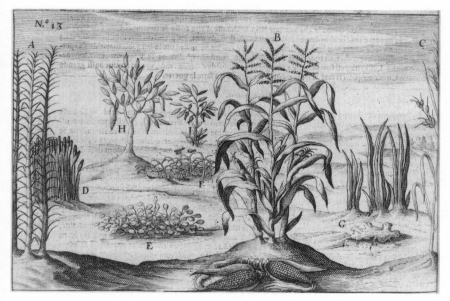

5. First image of maize in Africa, ca. 1602.

graving depicting distinctive plants of West Africa shows maize ("le Mays ou Blé de Turquie") front and center, in the midst of sugar cane, rice, millet, beans, peppers, and various spices.

In southern Africa, however, maize still may not have arrived on the Cape by 1652; Jan Van Riebeeck of the Dutch East India Company did not report seeing it there in that year, despite his hope of identifying new local food sources for the company's fledgling settlement. In fact, as part of his plan to provision Dutch East India Company shipping, in 1658 he asked for maize seed to be sent from home, to test its value at his planned supply station at the foot of Table Mountain.[8]

Searching for the initial introduction of maize is amusing but beside the central point—namely, how to understand and gauge early African reactions to it as a food and as a crop. In Africa as a whole, maize took its place in farm plots whose cultivation was already finely tuned to the vagaries of Africa's capricious climate and old soils. The new crop nonetheless held forth the promise of additional nutritional capacity that could stimulate population growth and concentration after its adoption. The arrival of this New World stranger offered benefits that were immediately obvious to farmers: traditional grain yields in Africa, less than half those of Asia and Latin America, reflected the obstacles of aridity and poor fertility that pertain in many parts of Africa. Maize and other New World crops (such as cassava, cocoyams, and potatoes) helped redress that imbalance in key parts of the continent. Unlike New World farms, which depended heavily on maize as a primary starchy staple, most African farmers adopted maize initially as a vegetable niche crop tucked within a complex system that relied on intercropping, rotation, and swidden (slash-and-burn) management of fertility. It is interesting to note, however, that while European observers first compared maize to chickpeas, Africans who named it tended to compare it to sorghum, choosing apparently to focus on the appearance of the plant itself rather than the shape of the kernel. The sorghum plant is almost indistinguishable in the initial stages of growth from maize, even to the eye of an expert. In the systems shaped by Africa's sharply defined wet-dry seasonality, averting risk required adopting a diversity of cropping strategies, rather than rewarding efficiencies of scale. It was the late (twentieth-century) shift maize made to being a monocrop grain staple that historically changed its role in the African diet and transformed modern African farming systems.

Genetic Expressions

For most African societies and agrarian systems we have no direct testimony about a historical "moment" that marks the advent of maize in local diets and in farmers' fields. Yet the hidden genetic and phenotypic evidence within the maize ear itself offers intriguing glimpses into the crop's historical geography. Though all maize types can cross-pollinate, each of the many cultivars has distinctive characteristics as regards its starch content, number of rows, color, plant height, resistance to insects, storability, maturity dates, and so on. These traits formed the smorgasbord from which farmers and consumers from year to year chose the taste, appearance, and growth characteristics they wanted to breed into the next year's seed. And of course each of the five major maize types (pop, floury, flint, dent, and sweet) moved into Old World settings after having already evolved in response to distinctive New World ecologies, consumer choices, and farming techniques.

The political geography of Europeans' New World conquests and transatlantic trade determined the patterns of maize's genetic emigration to Africa. Spanish ships, for example, began their New World contacts in the Caribbean and necessarily first encountered distinctive Caribbean flint maize types, identifiable by their hard starch, early maturation, and variegated grains in bright colors, especially red. Spanish flint maize first came to the Old World through Seville, from there to Venice, and then on to Africa via Egypt and the Nile Valley, where many observers remarked on the kernels' striking red color.[9] Roland Portères argues that such Caribbean flint maize then moved south and west across to West Africa following routes of the pilgrimage. Certainly, the core names of *makka* (Mecca) and *masa(r)* (Egypt)—in Fula, Hausa, and other West African languages—imply this route. At an early date, the Portuguese may have also introduced *cateto*, a yellow to orange flint maize from the uplands of Brazil and Argentina (called *Maíz argentino* in Cuba), a variety also found in the nineteenth century among Xhosa and Zulu maize farmers at the Cape and Natal respectively, but also with Bantu speakers who settled in southern Somalia, and in China. All of these were, of course, along lines of trade within the early Portuguese trade hegemony, though the evidence of their involvement is only circumstantial at best.[10]

Flint maize adapted well within many of the same niches in which Africa's indigenous sorghums and millets had thrived. Maize never fully replaced sorghum and millets in the drier areas of Senegambia, but it complemented them. In many other less arid areas maize quickly came to predominate and replaced local crops. The new crop's quick maturation and the tastiness of its roasted green ears also offered a low-labor food source in the "hungry season" preceding the harvest—in other words, well before slow-maturing sorghum was edible. In more humid areas of West Africa, especially the Upper Guinea coast, where rice was the dominant cereal, maize remained a niche vegetable, consumed fresh and neither stored or hand-milled.[11] Whereas rice dominated the moist bottomlands of that zone, maize could occupy uplands and serve as a pioneer crop in new forest clearings, a role it also played during its spread into China and Southeast Asia.

In other non-rice-growing areas of West Africa, farmers apparently disdained the hard starch and lower yield of flints and chose floury maize as their dominant staple, one that would replace or overshadow the flint types which complemented the agronomic cycles of rice zones so well.[12] Floury maize may have come later to Africa, since its origins in the Andes and dry areas of northern Mexico made it less accessible to early European settlers and merchants than maize types available from the New World's coastal regions and islands. Once it arrived, however, the soft-starch floury maize made dramatic inroads into West African areas, as one outgrowth of the forest-clearing and state-building efforts that took place on a large scale during the slave trade and the development of the Atlantic economy. The floury maize that dominated much of West Africa until the late twentieth century was served not as a vegetable, but as a grain to be paired with such New World root crops as cassava and local yams. Floury maize provided the essential carbohydrates to support large concentrations of population, which made possible the establishment of state power.

Floury maize, like its flint cousin, offered new options to West African farmers who needed an annual crop that could adjust to a forest-savanna mosaic, thrive in newly cleared forest soils, and prove transportable, divisible (for taxation purposes), and resistant to the preharvest bird damage that plagued sorghum, millet, and rice. In West Africa, as in northern China, maize turned out to be an excellent "relay" crop for cowpeas (or wheat, in China), to be planted between rows that matured in succession, and candidate for intercropping with other food crops, especially nitrogen-fixing legumes like beans and peas that added to the soil the nitrogen that maize plants craved. Most important, maize was the pioneer crop par excellence, in that it helped establish forest fallow cultivation systems in new frontier forest settlements.[13]

Outside West Africa, the movement and adoption of maize had a different history. Maize appeared in most areas of Africa within a century, or even a generation, of the birth of the Atlantic economy, but often unobtrusively as a novelty or niche crop. Most farmers in the Ethiopian highlands, for instance, kept it in their repertoire as a garden vegetable crop but until well into the twentieth century disdained its use as a field crop.[14]

In eastern Africa the introduction of maize followed closely on the heels of mercantile imperialism and then formal colonialism, though lexical information intriguingly suggests much earlier adoption. Along the Swahili coast, sixteenth-century Portuguese settlers may have begun cultivating maize to provision their garrison at Mombasa, and it became one of the dietary staples cultivated around those settlements. It is probable that maize also arrived via the lateen-sailed dhows of the Arab and Banyan (Indian) Indian Ocean trade, which probably also brought it to the Horn of Africa. In any case, by the mid-nineteenth century, it is clear, Swahili caravan towns that helped coastal traders establish the caravan route penetrating the Great Lakes region used maize as a food source, once entrepreneurial farmers along the way adopted maize as a salable commodity. Maize appeared as part of African farm repertories described in early European accounts of local farming systems generally dominated by starchy bananas *(matooke)*, New World root crops (including cassava and sweet potatoes), and African cereals (sorghums, eleusine, and bulrush millet). In Uganda maize appeared by 1860 in the garden agriculture of most of the major state systems—Buganda, Bunyoro, Toro, Ankole, and Acholi—where maize as a low-labor annual crop complemented perennial banana cultivation and local horticultural traditions. By 1876 Henry Stanley found, on the shores of Lake George in western Uganda, local cultivation of "an abundance of Indian corn, millet, sweet potatoes, bananas, and sugar cane."[15] By 1880 Emin Pasha (a European entrepreneur turned potentate) reported that maize was widely cultivated in the Acholi region of Uganda, where he had undertaken to introduce a "horse-tooth" maize variety that "thrives well."[16] He was probably referring to what Angolans call *dente de cavalho,* an early flat white dent found in southern Africa. Other Europeans dutifully recorded the presence of maize, and consumed it in their travels, often gratefully, though it was African farmers that found ways to build it into their diet and the rhythms of labor and daily life.

Cultural Expressions

Additional enigmatic evidence of maize's arrival in Africa is supplied by the cultural, aesthetic, and popular responses of African peoples who appropriated the New World crop into their diets and economic lives. West African societies adapted maize imagery in elements of ritual, popular oral tradition, sumptuary law, and material culture in ways that give some idea of its influence. In Akan oral tradition, one *ntoro* (exogamous patrilineal division), Bosommuru, has maize as its totem. The *ntoro's* day of veneration on Tuesdays features a ban on consuming maize on that day. In historian Ivor Wilks's reconstruction of oral traditions, Asante etiology reaches back to the late fifteenth or early sixteenth century as the origins of Asante cultural identity, the approximate time of maize's first arrival.[17] By the eighteenth century maize had become strongly associated with the Asante army and perhaps with state power in general. One Akan proverb notes, "Aduane panin ne aburoo" ("The chief [or elder] among foods is maize").[18]

Elsewhere in West Africa, maize appears as an anthropomorphic figure. In a nineteenth-century Yoruba folk tale, maize takes the form of a mysterious ghostlike stranger. One verse intones:

> I do not know at all, but I wish I knew
> What clothes the maize plant wore
> When he first came
> To this Origbo from Olufe town,
> But this I know surely
> That he was disguised upon his journey here.[19]

In Yoruba the symbolism of maize *(àgbàdo)* also plays a role in the Ifá oral divination verse called ÉsÉ Ifá. Babalao Wande Abimbola published a colorful translation of one example.[20] In my shortened adaptation of the oral text that he recounted to me, maize is assigned a gendered metaphor that celebrates the plant's almost human fecundity:

> All naked, Maize went to the farm, and she came back
> with two hundred dresses—
> Two hundred dresses!

All alone, Maize went to the farm, and she came back with
two hundred children—
Two hundred children!

This expression of wonder at maize's productivity is reminiscent of
sixteenth-century European amazement as related in Du Bartas's
Divine Weekes and Workes:

> Heer of one grain of maiz, a reed doth spring
> That thrice a year five hundred grains doth bring.[21]
> Du Bartas (1544–1590)

Maize also established itself within West African aesthetic traditions of material culture. In ceramicware, artisans among the
Yoruba and several other groups used discarded maize cobs as
a tool to inscribe a "maize cob roulette," a distinctive rolled de-

6. West African maize cob roulette.

sign, on pottery. At Ife, the ritual center of Yorubaland, potsherds with this design formed the paving materials in walkways for elite households. Common traditions among the Fon, Aja, and Nago peoples living in the historical state of Dahomey recorded that their kings reserved millet for royal consumption, in an effort to save that traditional crop from the rapid incursion of maize.[22] The full range of these rich and creative expressions, however, awaits collection by scholars of local traditions.

An even more fascinating, if often perplexing, measure of local historical perceptions of the new crop and the timing of its arrival locally is the rich panoply of names that African languages attach to maize. Comparing words for maize in African languages is a unique source, but one whose methods are intuitive and sometimes idiosyncratic. The rewards, however, are considerable, since collecting and analyzing the names of any new phenomenon can reveal aspects of the local imagination not evident in other sources. A number of patterns of naming repeat across linguistic groups, ecologies, and economic categories. Also evident are several kinds of encounters, universes of material life embodied in food, farming systems, and networks of cultural exchange. As with trade networks and ecology, these lexical networks sometimes cut across—and sometimes go against the grain of—formal language categories.

Naming the stranger was, in each new society that encountered maize, a process of trying to make the exotic more familiar. The most common Old World pattern for naming the new arrival was to combine the name of an already known grain with a description of its provenance (as popularly conceived). The perception of maize as a kind of sorghum is perhaps the most common African naming convention. Among East African language groups (Chagga, Akamba, Samburu, Pokomo, Taita) that must also have received maize first from the coast, the common name for maize is *pemba* (also cognates *bemba* in Kikuyu and *hemba* in Chagga). A. C. A. Wright argues that the word derived from the name of the island near Zanzibar on which Portuguese planters in the sixteenth century began to raise foodstuffs, including maize, to supply their coastal garrison. Historian Steven Feierman, however, says that the Shambai term *pemba* refers to sorghum, and Christopher Ehret, who has mastered the historical linguistic evidence, agrees that *pemba* is a reference to sorghum.[23] This gloss is made more plausible by an older term in Kiswahili, *pemba muhindi,* or Indian sorghum.

The second part of the pattern is the reference to a distant place, or allusion to the grain's coming "from the sea." In highland Ethiopia, Semitic speakers called corn *yabaher mashela* (Amharic), or *mashela baheri* (Tigrinya), both terms meaning, literally, "sorghum [from] the sea." In Malawi the Chichewa language called maize *chimanga* (from the coast), indicating an understanding of its origin resembling the perception of Semitic speakers in Ethiopia; on the East African coast the Kiswahili short-hand term is *muhindi* ([grain] of India). It is worth noting that the only African locality to use *burro* or *aburro,* borrowed terms derived from the Portuguese *milho zaburro,* was an early site of Portuguese trade—the earliest permanent Portuguese trading station on the West African coast, at El Mina (established in 1482) on the Gold Coast, where the young Christopher Columbus had first practiced his Atlantic navigation. In the Akan family of languages of the Gold Coast, *aburo* is maize, but Akan speakers also describe overseas countries as *aburokyire* (countries where maize comes from). The Akan call the English (their colonizers) *abrofo* (sing. *Oburoni*).[24] Historian of the Dutch Gold Coast James LaFleur says that *aburro* in the old Dutch transliterations is a direct reference to sorghum.

Near the mouth of the Congo River, Kikongo speakers in the mid-sixteenth century called maize *maza mamputo* (grain of the white man); Mande speakers in Senegambia offered *tuba-nyo* (white man's grain), a similar gloss. From Egypt along the trade route south to Lake Chad, local lexical terms for maize, especially in Hausa and dialects of Fulfulde, derived from the root *masa,* or *masar* (that is, Egypt), describing maize's likely path of introduction to the region. Among other groups, the Bambara use either *maka,* a word that suggests that the grain was introduced by pilgrims returning from the *hajj* (pilgrimage), or *kaba,* a word that also designates sorghum.[25]

Maize in southern Africa followed a somewhat different chronology of contact, though as in other regions the crop percolated onto African farms through international trade and population movement. Portuguese trade seems to have provided the earliest introduction, as suggested by use of names such as *zaburro* (in Mozambique, from the Portuguese *milho zaburro*), *masa mamputo* (discussed earlier), and *mealie* (Afrikaans, from Portuguese *milho*), a term now widely used generally by many linguistic cultures in southern Africa.[26] Other names have more enigmatic referents or have borrowed from neighboring areas where the crop had had a longer history. Swazi traditions, for example, associate maize's arrival with the origins of their Dlamini royal clan, but their siSwati language also borrows the Zulu term *m'lungu* (white man), or *u'hlanza-gazaan.*[27] Linguist David Gough of Rhodes University has used the most common Xhosa and Zulu words for maize (*umbona* and *umbila,* respectively) to measure early migrations and exchange between those two groups. Among the northern Tswana and Sotho, maize arrived with missionaries, who established agricultural stations as part of mission churches. The Sotho called maize *poone,* a version of the Xhosa term, while the Tswana called it *mmedi.*[28] Up to the first third of the twentieth century, however, southern African diets still consisted of sorghum and millet, reflecting landscapes where maize was sometimes known but never dominant.

Still other patterns of naming suggest local farmers' sometimes whimsical response to the physical appearance of the plant. In some Tigrinya-speaking areas of northern Ethiopia, the name for maize comes from what must have seemed to farmers the unusual feature of husked ears; they called it *ifun* (covered) or *ilbo* (uncircumcised). Other societies, like the Akan and the Yoruba, assigned images of sexual potency to the ears that bore maize kernels.

These patterns emerge most coherently when we look at particular language families and language regions. Sir Harry Johnston's impressive 1922 compendium of Bantu vocabulary contains a full list of words meaning "maize." Christropher Ehret has argued in some of his early work that Bantu languages that emerged from a root crop–"planting" agriculture over the span of about a thousand years of interactions with central Sudanic groups gradually borrowed and adapted the concepts of grain cultivation and processing.[29] The process by which Bantu speakers named the fairly recently arrived crop of maize seems to have been part of a longer-term adjustment to grain agriculture as a whole. Maize terms adopted as borrowings for the concepts of grain and grain processing in eastern Bantu Languages include: *-bele (grain), *-caka (sorghum), and *-papa (porridge), among others. The arrival of maize on the African scene after 1500 thus added a new layer to these older patterns.

Words for maize in the Bantu languages include virtually the full range of naming patterns seen in African languages as a whole. The term (pa) manga (of the sea) or mapira manga (sorghum of the sea) names maize in the Yao, Nyanja-Chewa, and Karanja languages. The term pemba muhindi (Indian sorghum) for maize appears in Kiswahili, Pokomo, Pare, Chaga, Thaiicu, Saghala, Seuta, Zalamo, and Kami. In several of these cases the languages have adopted a shorthand expression—either pemba or muhindi—similar to the American English shortening of the original term "Indian corn" to simply "corn" to mean "maize." In many other languages names for maize were simple adoptions of older terms for sorghum, such as saka (Sabi, Fipa, Mambwe, Lungu, Nyiha, and Bambusangu), bila or pila (northern and southern Nguni, southern Sotho, Holengwe), tama (Gogo, Kagula, Sagala, and Ngazija), or kusa (Konjo, Amba, Littuku, Vamba, and Bira). In some cases languages chose to adopt their maize name from bulrush (or pearl) millet, an

African crop generally confined to the driest areas, where maize could not survive. Thus for Hehe, Bena, Kinga, Lenje, and Nsanga, the word *cebele* (or *kebele*) meant maize, while for other groups in more arid zones *cebele / kebele* meant the indigenous bulrush millet more familiar to them.

At least one more common pattern for naming maize appears in a number of Bantu languages. That is, the identification of the maize plant not with other grains, but by the appearance of its stalk. So, in Nyamwezi the word *dege* means "maize," but in Southern Cushitic, from which they borrowed the word, the term *dig* translates as "stalk." The widespread Bantu term *konde,* meaning "thick stalk," is the word for maize in a cluster of languages of the region. The word *gombe,* which means "maize" in Nkoya, means more precisely "thick stem" or "stick" in a scattering of other languages. In a wide range of savanna Bantu languages (Ovimbundu, Bemisa, Yao, Zulu, etc.) *ngoma* denotes a heavy stick, whereas in Bisa, Lala, and Nsenga it means "maize."[30] Often the Bantu maize word list indicates that some farmers named the maize plant by a word meaning "stick" or "stalk." Perhaps the practice was especially common among language groups that had historically relied on root crops or "planting" agricultural systems.

Finally, exactly who was it that named the new plant and its distinctive food products? In historical and prehistorical eras that predate the arrival of maize, Christopher Ehret argues, many non-grain-planting cultures simply borrowed words from languages of the central Sudan, where a tradition of cereal cultivation had provided them with a rich vocabulary for grains and boiled grains (porridges).[31] It may also have been women who either borrowed or coined words to apply to the arrival from the New World. Certainly, in the modern era women make up the majority of Africa's farmers. In the twentieth century, as maize shifted from a household garden crop to a valuable cash crop, it often became a male domain. We know from more recent evidence that farmers apply names of their own invention to new maize varieties according to much the same patterns the original namers of the new plant followed. Ample evidence exists to show that in recent times women were the namers; for example, in western Kenya, Jean Hay tells us, it was women who named crop varieties after male labor left in early migrations to the cities. In 1917 a new type of white dent maize was novel enough that Luo women called it *orobi* (for Nairobi, founded in 1901)—perhaps a sign that it marked a new era in their economic lives. When maize types, in the modern parlance for hybrids and genetic improvements produced by national seed centers and private corporations, merely take the numbers and initials of scientific trials or the location of the research station (such as Bako Hybrid 660 or Kitale 522), farmers no longer take the initiative to imagine the plant's characteristics. Thus the bland and homogenous comes to be preferred over local aesthetic expression. By now, maize has completed its acculturation to Africa and is no longer the stranger.

Maize's Invention
in West Africa

3

Was the arrival of maize in Africa a watershed
event in defining either the historical direction
for the continent or particular historical set-
tings? Some have made the argument in a general way that it was,
and others have made a quite specific case that maize was a distinctive catalyst. What is the evidence for such determinism in Africa's
encounter with the New World crop?

Environmental geographer (and gastroenterologist) Jared Diamond has argued—persuasively, some would say—for environmental determinism over the long haul in human history. He suggests that the global distribution of natural resources, particularly
in food supply, set in train the long-term development of the human economic and cultural landscape. He notes that while human
intelligence and ingenuity were everywhere in evidence in world
history, no level playing field existed when it came to the global distribution of environmental raw materials: not all areas of the globe
were equally endowed with natural resources such as animal and
plant germplasms.

Even though Diamond's focus was global, his thesis about the in-
equality in endowment has specific implications for Africa. In the
case of genetic types, for example, some regions of the world had a
rich pool of domesticable livestock species, while others did not.
Africa had only the donkey as a potential domesticate (and that

only in the Nile Valley) and had to wait for cattle, sheep, goats, and camels to arrive from elsewhere (the New World also had to get along without these until the Columbian exchange of the fifteenth and sixteenth centuries).

The potential genetic resources for agriculture in Africa were also unbalanced. Of large-seeded grass species the Mediterranean world had thirty-two types that were potentially domesticable as cereal crops, the Americas had eleven (including, of course, maize), while Africa had only four, none of which would be one of the world's primary food grains of the twentieth century—wheat, barley, rice, or maize. Most of the genetic stock of grain that has become global domesticates was distributed in the Northern Hemisphere and generally tended to spread along an east-west axis, rather than from north to south.[1] In this view, therefore, Africa had to overcome an early liability, which, of course, it eventually did by adapting exotic crops, once they became available via trade, accident, and human design, to particular African ecological settings. But this process of appropriating key New World food sources in Africa took some time, and it began only half a millennium ago. Diamond's schema is global in its sweep and not focused on specific crops; we nevertheless can ask the question How central was maize as a food source to the formation of particular types of political and social systems in the Old World, once it finally arrived?

While many or even most historians scrupulously avoid explanations based on a single cause, maize has tempted a number of writers to ascribe great power to it. The noted scholar of world history William McNeill, for example, has attributed the demographic and political domination of particular ethnic groups in the Balkans to their incorporation of maize into their agricultural and food systems:

When Greeks, Serbs, and Vlachs found that the new maize crop allowed them to live all year round in the high mountain valleys, where they were safe from the twin scourges of the plains—malaria and Turkish oppression—the political and economic balance of the Balkans began to shift . . . Thus one may say that what potatoes did for Germany and Russia between 1700 and 1914, maize did for the mountain Greeks and Serbs in the same period of time. In each case, a new and far more productive food resource allowed population to surpass older limits, and larger populations in turn provided the basis for the enhanced political and military power attained by the four peoples involved.[2]

Alfred Crosby, another thoughtful historian on the grand scale, stops short of linking the appearance of maize with the rise in southern Europe's population and political fortunes, but he nonetheless traces the crop's central role as a new source of calories in Romania, Hungary, the Po Valley, the Danube Valley, and the Caucasus. Crosby, like McNeill, turns the assertions of that persistent pessimist Thomas Malthus on their head by arguing that the effects of new crops (such as New World maize) on food supply accounts for population increases and economic growth in the early modern period, although he does hedge his bets more than McNeill:

An entirely new food plant or set of food plants will permit the utilization of soils and seasons which have previously gone to waste, thus causing a real jump in food production and, therefore, in population. But, before we accept this statement as gospel, let us acknowledge that we are taking much for granted. How can we be *sure* that a population which simultaneously switches from wheat to maize and increases in size could not have accomplished the same increase without having heard of maize? Perhaps the switch to maize came not because of its greater productivity but because the people in question simply liked the way it tasted. Perhaps the increase in population stemmed from a dozen or a hundred factors having nothing to do with maize.[3]

According to both Crosby and McNeill, the scale and nature of the changes wrought by maize were profound in parts of the Old

World, but were there similar effects in Africa? And were such transformations universal in state and social systems there?

The penetration of maize into West Africa took place in many cultural, economic, political, and ecological settings. Peoples who settled around the Bight of Biafra, the Bight of Benin (both in present-day Nigeria), the historical kingdom of Dahomey, the Gold Coast, and the Ivory Coast accepted maize with great alacrity. But the reception was far less enthusiastic in Guinea, Sierra Leone, Senegal, and Gambia (the Upper Guinea coast), where endemic African varieties of rice remained rather firmly entrenched. The factors influencing the warmth of the local welcome for maize were a product of a grab bag of conditions: local ecology, the types of maize that presented themselves, local farming systems, crop niches, and even the aesthetic sensibilities of African consumers.

To examine these factors more closely, we will look at the case of Asante in West Africa in the centuries immediately following the Columbian exchange and follow the factors into the modern period. Maize's performance, alongside other New World crops, was a catalyst in the human transformation of the forest in Upper Guinea that stretches from Sierra Leone to the Gold Coast and the Dahomey gap, then on to Nigeria. This ecological zone, which mixed humid forest with savanna and landscapes once called derived savanna, but more accurately called forest-savanna mosaic, was the home base of major state systems that emerged in the seventeenth and eighteenth centuries.[4] These states included kingdoms, memorable for their great military, political, and artistic achievements, that straddled forest and savanna ecozones— Asante, Dahomey, Benin, Oyo.

Statecraft in the Forest: Asante

Did all African societies that welcomed maize become winners, the beneficiaries of maize's grace, or blessing? The Akan region of the Gold Coast seems to have been such a place, where ecology, local politics, and economic response to the expanding Atlantic economy joined together to generate the powerful and long-lived Asante Empire, beginning at the outset of the eighteenth century. Asante offers a particularly powerful case for the importance of maize penetration, for the empire can boast a rich set of historical sources and an insightful tradition of scholarly writing about its history, as well as a highly visible tradition of producing textile (such as *kente* cloth), metal objects (brass goldweights), and political institutions (the Asantehene, the queen mother, and their courts). The origins and timing of Asante's rise to power in the seventeenth and eighteenth centuries has presented a challenge to historians. The product of both internal dynamics and the opening of the trade links to the Atlantic world, Asante's rise and the introduction of maize thus offer an intriguing case for assessing maize's historical role in a humid and semihumid West African political and ecological setting and provides broad insights into the agroecology of West African political history as a whole.[5] Maize was part of the historical conjuncture that resulted in Asante's historical prosperity and hegemonic growth—less as a causal factor than as a necessary ingredient in the historical process of population and political growth, in which food was a critical factor.

In the forest ecology of Upper Guinea, the primary food-related dilemma affecting human settlement and state-building was not a lack of protein, but the paucity of carbohydrates. Wild yams and other tubers were part of the forest's biodiversity but existed too sparsely on the ground to provide a caloric base for an army, a bureaucracy, a population of town dwellers, or nucleated villages of taxpayers. While lowland rice *(Oryza glaberrima)* was endemic to the northern Upper Guinea forest in Sierra Leone and Guinea, it seems not to have made an impact on the drier, semideciduous forests of central Ghana and the historical Gold Coast.[6] Other African grains, such as sorghum and millet, which sustained most dense savanna settlements, were long-maturing, needing the sun and a long dry season to ripen into a viable food source. Cultivated yams, rich in carbohydrates, were well adapted to shady plots and forest soils but were also a long-maturing, labor-intensive root crop that was more a prestige food than a reliable staple.[7] Forest soils were fertile and moist on surface levels where leaf debris decomposed, but sunlight, the forest's most precious commodity, rarely reached the understory plants on the floor of the closed-canopy forest. The tallest trees won the competition for sunlight, allowing only shade-loving plants to occupy the forest floor in a mature forest. Moreover, forest soils, once cultivated, were easily leached of nutrients. Those soils then required subtle management to produce food in large amounts.

Solving the historical puzzle of the foundations of forest statecraft and population density requires a fairly bold vision, given the paucity of archaeological and historical evidence available. Historian Ivor Wilks, using tantalizing shreds of historical evidence and the botanical facts of forest ecology and crop agronomy, took such an approach. In a 1978 study Wilks linked the Akan people's historical evolution of a distinctive (forest) fallow agriculture with the emergence of the Atlantic economy. To this convincing but still speculative formula, we need to add maize as a defining ingredient.

Fundamentally, the constraints on human settlement of forests were 1) how to remove primary, high-canopy forest vegetation to allow sunlight to penetrate to food crops at ground level, 2) how to prevent vegetative regrowth from choking fields after clearance, 3) and how to mix crops to provide a sustainable food supply that did not degrade forest soils. The figures for premodern tropical forest clearance are staggering: clearing a single hectare of primary tropical forest required removing 1,250 tons of moist vegetation by using cutlasses, billhooks, and fire. It was hot, dangerous, arduous work to remove the "cumbersome growth of fibrous stems and vines, mixed with other plants of a watery nature."[8]

The real surprise in Wilks's equation, however, was not so much the daunting weight of biomass to be hacked, cut, uprooted, dragged away, and burned, but the fact that after the first clearance of 1,250 tons, subsequent clearance of the same plots after a fifteen-year fallow was only 100 tons! Thus, the key task had been the initial breaking of the forest canopy's monopoly on sunlight, through the first-stage removal of both the canopy and the choking understory of trees, bushes, and vines. Once the forest's primary canopy had been cleared, an agricultural economy could expand the frontier of cultivation into the forest. In later stages, labor could be released for crop cultivation, military service, construction, and so forth.

Nineteenth-century observers tell us that this ecological revolution had already taken place by the time of their arrival. At the time of Joseph Dupuis's visit in 1820 to the Asante heartland, the forest fallow system was well entrenched and much of the land clear-felled, leaving a landscape akin to "the country gardens of Europe."[9] How was this cleared and cultivated landscape achieved, given the high initial labor costs of clearance?

Wilks answers the labor question through intriguing oral and documentary evidence that points to the fifteenth and sixteenth centuries as the time of a crucial conjuncture that drew new human populations into the central forest and provided the social mechanisms (matriclans or *abusua*) to integrate them. The first European contacts with the West African coast in the fifteenth century in search of gold found an already active slave trade that brought captive labor from the Niger delta to the Ghanaian coast. As the Portuguese traders quickly learned, the easiest way to obtain gold from the mines of central Ghana was to transport slaves from elsewhere on the West African coast to the new Atlantic entrepôt (founded in 1482) at El Mina. Incorporating these new populations as slaves, fictive kin, and dependents provided both the labor and the mechanisms of social coercion that permitted state systems to evolve.

One ingredient hidden in this story points directly to the year 1500 as a takeoff point for changes in the ecosystem: the arrival from forest ecosystems of Central and South America of new food crops ideally suited to feeding expanding forest-based polities. New World food crops—cassava (manioc), cocoyam *(Xanthosoma sagittifolium)*, and above all, maize—brought by European ships to provision their coastal fortresses and island plantations, spread quickly beyond the fortress gardens via local trade networks. The infusion of these plants into the local agronomic ecology spurred an agricultural carbohydrate revolution that allowed forest peoples of the Upper Guinea to feed a dense, growing population and fostered an elite political class, royal courts, and a standing army. Maize, and cassava, were the biotic wedge of a human assault on the forest landscape to convert the forest's biomass and energy into usable carbohydrate calories.

Residents of the Gold Coast (speakers of Akan languages) who cultivated these new crops clearly understood their importance from the beginning and probably were among the first African societies to recognize their value. The Akan word for maize, *aburoo,* which connotes "foreign," has no wider usage in West Africa and does not seem to have been borrowed from any neighboring group.[10] In fact, speakers whose first language is Akan can often provide no gloss for the term *aburoo,* though it appears in older Dutch documents in a way that clearly indicates that it meant something like "sorghum of the foreigner," the most common gloss throughout Africa.[11]

The formal historical sources are vague regarding dates for the arrival of maize in West Africa and the actual points of entry. Evidence of the profound nature of its economic and cultural impact comes primarily from local representations of the plant's deep cultural symbolism in personal and public ritual. It also comes from ecological inferences about how one branch of the Akan-speaking people came to transform a thinly settled forest environment into a powerful, expansive empire that fed and otherwise sustained a sophisticated state bureaucracy and military.

The primary weapon was a local forest fallow repertoire, a historically accumulated body of local knowledge combined with revolutionary plant germplasms. Observers of the forest fallow farming system described elements of it as early as the first part of the nineteenth century, but its fullest elaboration appears in the insightful fieldwork of anthropologist Kojo Amanor. What Amanor, working in the 1990s, describes as a twentieth-century adaptation, historical documents and data from other forest zones suggest was in fact a historical process that evolved over time and was the system (or a set of related localized practices) in place throughout central Ghana after the sixteenth century, when forest farmers and gatherers added maize and cassava to specialized niches within forest cultivation.[12] Similar processes probably took place all along the Upper Guinea coast and into the coastal area of the Bight of Benin.

Ghana's Upper Guinea forest cultivation system—forest fallow—derives its rhythms from two seasons of rainfall and a season of dry wind (the harmattan) that blows dry air and dust from the Sahel toward the Gulf of Guinea. The heavy rains take place between March and early July; a second, smaller rainy season beginning in September and lasting until October allows the cultivation of a second plot. The farm cycle began with the clearing of fields for the primary farm between December and January. Farmers use cutlasses to slash and remove the understory of shrubs and ground cover. Farmers then cut major branches from large trees (a technique called pollarding) to open the canopy cover and added leaf debris as a mulch to the soil's surface.

With the beginning of rains in March farmers placed leaf debris, smaller branches, and other now dried vegetation in piles for controlled burning. Fire reduced forest biomass to ash and destroyed insects and small weeds. Burning also released phosphorus and singed the leaves and branches that fell to the ground, thus adding organic matter to soil, enhancing the layer of mulch, and increasing the penetration of sunlight to the forest floor. Clearing, burning, and pollarding, in effect, converted the energy stored aboveground in vegetation into soil nutrients to feed nitrogen-loving food crops such as maize.

During the March rains farmers also planted yams in mounds

near small trees that served as stakes for the plant's emerging ten-drils. Once the March rains were in full swing, farmers planted maize, using minimum tillage—a technique now much in vogue among development agencies—to keep mulch and soil moisture in place. The maize thus received the moisture essential for its early growth and tasseling. Two weeks later, farmers planted cassava sticks between the germinated maize plants, and cocoyam corms preserved in the soil began to sprout. As these crops grew, dense vegetation emerged, with maize leaves leading the way toward the intense sunlight they require. The leaves of the young cassava and cocoyams closer to the forest floor cooled the maize roots and pro-tected the forest soils from the impact of rainfall and direct sun.

Understanding the dynamics of the forest fallow system's impact on food supply requires that we interpret against the historical background the role maize played in wider change. The capacity of maize to provide two harvests within a single season gave a strate-gic boost to local food supply. The availability of food, in turn, re-leased labor to extend the frontiers of forest settlement and support development of politics and statecraft.[13] The forest fallow crop rep-ertoire as a whole was thus part of an agricultural transformation that drew new labor into the forest and broke the knot of achieving that first clearance of the primary forest. One of the engines for the Upper Guinea forest fallow revolution was the arrival of three New World domestic plants that occupied strategic niches in forest culti-vation. As a new cultigen, maize offered an advantage: it was an early-maturing food source that provided carbohydrates by the end of the rains, while exacting less work than yams did. Maize also gave a second harvest, while cassava complemented maize's early yield and double crops by supplying another low-labor (but long-maturing) crop, which could remain stored in the ground for ex-tended periods. It is therefore not surprising that historical sources report that in the era of the slave trade maize increasingly domi-nated forest and coastal cultivation.[14]

The types of local maize found on forest and forest mosaic farms

of modern Ghana may well represent the kinds of maize that originally found their way into the forest agricultural system. In turn,
the maize types allow us to infer something about their role in
foodways and in farm practice. Flint maizes, such as the Brazilian
orange and yellow Cateto, would have matured early to supply a
food that could be boiled or roasted at the milky stage. The Cateto
race is the group of yellow to orange maize types occurring in
Brazil and assumed to have been grown by indigenous people living along the Atlantic coast from Brazil to Guyana.[15] Although not
a high-yielding type itself, Cateto readily crosses with many other
races of maize and therefore may have been the basic maize variety
that West African farmers adapted to their local needs, through
a technique that maize breeders call mass selection. Floury maize
is also a distinctive type found among local West African varieties. When hand-milled with a mortar and pestle, floury maize also
would have offered the raw material for kenkey, a polenta-like stiff
porridge steamed inside the cornhusk and eaten with palm oil and
vegetable or bush meat stews. Over the course of the late eighteenth and nineteenth centuries, maize continued to influence state
expansion and economic growth. First, it provided a transportable
forest-based food supply for Asante's army that expanded its reach
into savanna zones to the north. Second, farmers in the savanna,
probably using mass-selection techniques, adapted their floury
maize to the drier environments at the savanna's edge, where the
new crop slowly replaced sorghums and millets—a process that
replicated itself up and down the West African coast, in various local forms, into the modern era.

From Asante to the Gold Coast

By the middle of the twentieth century, Ghana's farmers had inge-
niously bred local maizes in early- and late-maturing varieties ideally adapted to their different farming systems and to the physical
environment. By the twentieth century four major zones of maize

cultivation had emerged, each reflecting its own ecological niche and place in the farming system:

Coastal savanna: Maize is intercropped with cassava and the drought risk is significant.

Forest zone: Maize is grown on scattered plots, usually as an intercrop with cassava, plantain, and cocoyam. Farmers plant most of their maize during major rains.

Transition zone between forest and savanna (forest mosaic): Farmers use intercropping but plant most of their maize as a monocrop, sowing the crop during both major and minor rainy seasons.

Savanna: Maize in this northern zone is increasingly important, replacing millet and sorghum. Farmers often grow maize in rotation or as a relay crop with groundnuts and cowpeas. Farmers cultivate many fields permanently and frequently fertilize their maize.[16]

This modern geography of maize cultivation has evolved from an older pattern that seems to have reflected two lines of entry for the crop. The first was via the coast and trade contacts with Portuguese, Dutch, English, and Swedish forts along the coast in the sixteenth and seventeenth centuries. Here the word *aburoo* seems to indicate direct contact with foreign traders. A second set of terms found in the northern part of present-day Ghana derive from the Hausa word for maize, *masar,* an association suggesting either a term borrowed from other groups in the West African savanna or introduced as an innovation by Hausa speakers whose contact with maize would have included travelers on the hajj returning via Egypt.

Maize cultivation in the British Gold Coast colony evolved in increments, as a part of an overall agricultural economy. One surge came in the 1940s, when the virulent swollen shoot disease sharply reduced cocoa plantings, thereby opening up new forest lands for maize as a cash crop.[17] In the early 1950s, southern rust (*Puccinia polysora,* or American rust) threatened the maize crop of the Gold Coast colony and West Africa as a whole, and consequently the urban food supply, but maize survived on farms and continued to play a still relatively minor role in the local and national diet.

In the years after independence, however, new transportation networks encouraged migration of farmers to savanna agricultural zones in the north, where in the late 1970s new efforts at agricultural extension and use of fertilizer inputs encouraged production of new maize types. Initially these programs met with some resistance. Farmers were concerned about inadequate husk cover in improved varieties, which invited insect and bird damage to maize in the field or in silos, when it was stored in the husk. Government maize breeders working at the Ghana Crop Research Institute began to develop new varieties based on farmers' preferences. The breeders sought to improve grain qualities and husk cover and to

7. Maize in the savanna, northern Ghana, 1975.

select for nitrogen efficiency and resistance to the parasitic weed *striga,* drought, and stem borers. In the late 1990s, Ghana's national government began an aggressive policy of promoting new maize types and use of chemical fertilizers and herbicides under the guidance of the Sasakawa Global 2000 program, which has promoted maize "package" programs in various parts of Africa. The results in yield and area planted were remarkable. From 1975 to 1995, the area planted in maize had more than doubled in Ghana, with yield increases of 50 percent; maize often replaced both sorghum and cocoa (Appendix Table 3.1).[18] By the end of the twentieth century, more than half Ghana's maize was produced in the north of the country as a monocrop, a far cry from its origins as the pioneer crop that promoted human settlement of the forest.

Whatever its origins, by the last decade of the twentieth century maize had established itself as an intrinsic part of the national diet, emerging as the most important cereal crop in Ghana, where it favorably complements other crops as the cheapest source of calories: in the south cassava, and in the north sorghum and millet. Nevertheless, unlike in southern and eastern African diets, in the average diet in Ghana maize contributes less than a fifth of the calories, whereas roots and tubers supply more than half the calories for rural households and almost two-fifths for urban households. Even where maize is the principle staple, as in the southern-central and Volta regions and parts of the northern region, it rarely contributes more than a third of the calories of the average household. Because the maize-based parts of Ghana's national cuisine require skill to prepare, most consumers do not purchase maize as grain, but acquire it in processed form. In the late 1980s well over half the maize growers also purchased prepared maize products at the market.[19] Prepared as kenkey, banku, or akple, maize has taken its place alongside cassava and yam as part of the national cuisine and farm economy.

The most significant recent development for maize in Ghana has been the introduction of quality protein maize (QPM), dubbed

Obatumpa (mother's milk) by the Ghana Crop Research Institute and released to farmers nationally as the primary new maize type. Ghana has thus become the first site in Africa for distribution of the new generation of maize, in what is likely to become a worldwide effort to promote the role of maize as humankind's principal food.[20] In Ghana's food supply the process has thus begun of shifting the status of maize from that of a vegetable crop mixed into the biologically diverse forest ecology to that of a homogeneous monocrop which increasingly dominates the national diet.

An Alternative Outcome in Nigeria

The historical experience of Asante, the Gold Coast, and Ghana is a regional variation on the larger maize story in West Africa, which differs substantially from that in eastern and southern Africa. Since the 1970s in western Africa, maize has achieved the highest rate of increase of any major food crop. Yet within West Africa and within particular cultural and ethnic clusters, the effects of maize have differed considerably with respect to world markets, local cultural traditions, and national economies. The Africanization of maize took place first in West Africa, in fields where farmers selected from among New World–bred maize types—or land races—for characteristics of yield, timing of maturity, and adaptation to both disease and climate. It also took place in women's market stalls, where consumers and sellers negotiated for corn with their preferred color, texture, and taste. Nigeria, which by the end of the twentieth century produced almost half the maize grown in West Africa, provides a useful contrast to Ghana's case. As in Ghana, in Nigeria the dominant agroecological zone shifted dramatically in the late twentieth century.

As was the case with the Asante empire, for Nigeria the initial encounter with maize came with the opening of the Atlantic system after 1500, though in ways that are rather surprising. Though maize never achieved the dominance over diet it would assume in eastern and southern Africa in the twentieth century, its history in West Africa, particularly in Nigeria and Dahomey (now Benin), suggests a profound impact. Presumably, the way maize fit into a niche within forest and forest mosaic farming systems resembled what happened in Asante. The appearance of maize in Yoruba Ifa oral traditions, however, also indicates a strong and long-standing engagement with maize as food and as social symbol. The core word for maize in the Benue-Congo language family, *agbado,* and variations recorded in the nineteenth century such as *agbade* (Dahomey), *egbado* (Daumu), *gbado* (Eki), or *gbadye* (Mahi), reflect the widespread cultural influence of the Yoruba culture and agricultural system and their history of links in all cultures of the Atlantic rim.[21]

In most cases the linguistic evidence in Nigeria follows patterns common elsewhere in Africa, including references to the putative provenance of maize and allusions to its resemblance to sorghum. In northern Nigeria the term for maize in Hausa *(masar),* is found in at least nine other local languages, an indication of either extensive borrowing or a common influence from migrants who brought the new crop. There are, however, some fascinating exceptions. Among the most colorful are the glosses in Piti (sorghum with a hat), Mala (sorghum that carries a child), Izere (musical horn sorghum), and Ikulu (sorghum like a *borass* palm sprout).[22]

Most historians and maize specialists have assumed that the main point of arrival for maize was the west and the Atlantic coast (as was the case in the Gold Coast evidence), probably via Portuguese and other European trade contacts with the southwestern Yoruba and Beni peoples.[23] The impressive lexical evidence assembled and analyzed by the Oxbridge group of Roger Blench, Kay Williamson, and Bruce Connell, however, offers a stunningly different view. They looked at both the places mentioned within maize names and the evidence of borrowing and the sound shifts within words. Virtually all the names for maize within the Nigerian panoply and the geography of their distribution refer directly or indirectly to a northern origin for maize. There were two means of diffusion, farmer-to-farmer exchanges and long-distance trade along an east-west axis (and sometimes from the northwest or northeast via the savanna empire of Bornu and the Niger River). The word *masar* and its variations suggest that the Bornu route that linked to the Nile Valley and the pilgrimage to Mecca via Egypt was perhaps the most significant. In southern areas the appearance of the *agbado* and *-kpa* clusters of terms for maize indicate arrival and diffusion from the northeast and also arrival from the west via the old area of Dahomey and expansion within that empire.

Surprisingly, the language evidence from Nigeria indicates only a

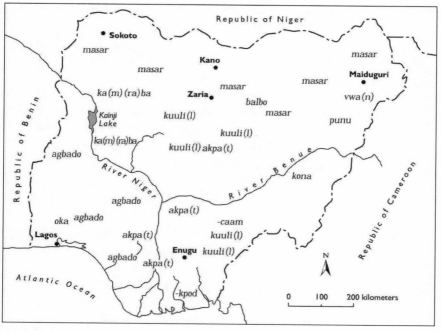

8. Maize terms in Nigerian languages.

minimal role for the Portuguese, and then only in the southwest, at the Bight of Benin. Only one language there, Isekiri, seems to use a word of Portuguese derivation, *imiyo,* a term that Blench, Williamson, and Connell link to the Portuguese *milho.*

The history of maize as a forest vegetable crop cultivated alongside root crops and legumes (including New World beans and squash or African cowpeas) dominated its early tenure in West Africa. But as in Asante and Ghana, the geography and agroecology of maize changed inexorably throughout the colonial period and during the final quarter of the twentieth century.

By the middle of the twentieth century, Nigerian farmers, processors, and consumers had established an elaborate repertoire of tastes and aesthetic preferences. Dutch maize researcher C. L. M. Van Eijnatten describes the intricate geography of preferences in color and type that had evolved in Nigeria by the mid-twentieth century:

> In northern Nigeria most of the maize cultivated is of the pale yellow colour and is flinty, although white flinty varieties do occur quite regularly. Around larger towns white floury varieties are regularly cultivated, but these only on a small scale and invariably only for sale to "southern" inhabitants of these towns. In the northern and eastern parts of this region the occurrence of dark brown, red or blue coloured grains is quite common.
>
> In the whole of southern Nigeria the people have a strong aversion towards brown, red, or blue grains. A definite preference for white colour does exist in Oyo, Abeokuta, and Ijebu provinces of Western Nigeria and in the eastern part of Eastern Nigeria, both of which produce mainly white varieties. It is in these areas that type and colour preferences are very pronounced, especially in the Yoruba communities of Western Nigeria. In most other areas colour preference is of little importance. Either of the two colours, white or yellow, will be acceptable, provided the grain type is suitable for the preparation of the foodstuffs. The preference for flinty

or floury grains is less pronounced than the colour preference even in the "white floury" areas of Western and Eastern Nigeria. When colour preferences were plotted on a map of Nigeria a mosaic is obtained, of intermingled preferences, varying from one town to the other. Outside of these areas flinty grains were normally judged more acceptable than soft, floury types, apart from Niger province. In Illorin province just north of the distribution area of white floury maize varieties, the white floury "Yoruba maize," as it is called, was definitely not acceptable.[24]

In the last quarter of the twentieth century, Nigerian and West African maize continued to evolve within local ecologies and national economies. In the early 1970s the humid forest and forest mosaic zones accounted for 60 percent of maize production, the savanna ecozone for only 25 percent of it. By 1983–1984, savanna ecology accounted for more than 50 percent and the humid forest and forest mosaic for just 23 percent. Most of this reversal is the result of growing improved varieties in monoculture.[25]

The release of a new maize type called TZB by the International Institute for Tropical Agriculture in Ibadan was a major breakthrough, as it offered an increase in yield to six times that of sorghum and millet and seven times that of cotton. In Nigeria's northern savanna zone, group interviews conducted by a research team in the late 1980s indicated that maize, considered a backyard crop in the mid-1970s, had emerged in 1989 as the lead crop in 90 to 100 percent of the villages surveyed in the northern Kaduna and Katsina districts. As in Ghana, maize offers a higher return on expenditure in land, labor, and cash than all competing crops except cotton (which remains a minor crop because of instability in the world market). Sorghum is only about a third as responsive to fertilizer as maize.

The trend in West Africa's maize economies to move from growing the crop in a biodiverse forest ecology to virtual monocropping in savanna agriculture may be an increasingly fragile one, as farmers calculate risks of their economic position. It is interesting that the economic collapse of Nigeria in the early 1990s that brought on structural adjustment programs from the International Monetary Fund and other international donors resulted in the loss in government import subsidies for fertilizer and an almost immediate decline in maize production. Farmers reverted quickly to sorghum and cassava in drier areas and to local rice in wetter zones.[26] This pattern stands in substantial contrast to that in two imperial systems which we will now examine.

Seeds of Subversion
in Two Peasant Empires

4

The initiation of sustained contact across the
Atlantic Ocean at the end of the fifteenth century set in motion a number of significant ex-
changes in politics as well as ecology, including the reframing of
cultural universes throughout the greater Atlantic world. The Mediterranean "lake" of the classical world quickly lost its pivotal position and became a part of the periphery, marginalized in both
economic and intellectual terms. Part of the fallout from that
change was a weltanschauung that had to accommodate the entire
New World, along with changes in material life wrought by the
transfer of new elements of political economy, by the foundation of
empires of extraction (that is, the forcible removal of natural and
human resources, from gold or spices to slaves), and by the exchange of organisms, which would find new ecological niches
when they crossed the Atlantic basin. The changes comprised a
conglomeration of new ideas, new economies of labor, and possibilities inherent in new foods, new crops, and their transformation
of Old World agrarian systems. If the struggles were human, the
weapons were often technological, and they embodied new possibilities for agriculture.

It is instructive to examine the ways in which two Old World
agrarian systems responded to the effects of *Zea mays,* the food
crop that had arrived as a deus ex machina at the opening of the

sixteenth century. In settings very different from West Africa, the *terraferma* territories of northeast Italy's Venetian Republic and the Christian kingdom located in the northern highlands of Ethiopia in northeast Africa came up with quite different responses. How did Old World peasant farmers react to the new food source, and how did overarching empire states set the terms of their response to it in the two critical centuries following the voyages of Columbus?

On the surface, these two agrarian empires had little in common. Venice had built a far-flung eastward-looking commercial empire, but in the sixteenth century the Venetian state turned its gaze back homeward. Venice sank new local roots by managing farm estates in its own agricultural hinterland in the valleys of the Po and Brento Rivers. The Ethiopian highlands relied on dryland farming, extractions, and the trade in exotic goods with the neighboring lowland zones to the south and west. These two imperial systems lay at opposite ends of the pre-Columbian Mediterranean world and reflected the economic and spiritual influence of a faded Roman trade hegemony and the intellectual legacy of Eastern Christendom.[1] While radically different in their ecology, forms of social property, commercial context, and political culture, the two systems nonetheless shared certain essential characteristics. Both had their economic base in smallholder agriculture and featured a social system in which peasant households played a significant role. Their adoption of maize offers an interesting comparative lens.

Venice is, of course, not an African case. Its inclusion here serves not only to broaden the basis of comparison, but to acknowledge that the Venetian trade diaspora was one of the primary means by which maize reached Africa, via the Nile Valley and the Red Sea. Venice's mercantile world intersected Africa's in cultural, economic, and religious spheres, as is evident in the *millefiori* glass beads traded in West African markets and recently unearthed in Aksumite tombs, in Venetian domestic ornamentation (such as doorknobs with African themes), and in the public architecture of the Nile Valley (obelisks) highly visible along the Grand Canal. A comparative link through maize enriches and illustrates connections of symbol and substance.

Examining the response to maize as a crop and potential food source within the elite and peasant cultures of these two imperial systems throws into sharp relief some fundamental historical truths about the social and economic change in agrarian systems in general. In this case the comparison focuses on conceptions of property; the relation of the state to rural production; ideal conceptions of family and inheritance; and the links between past patterns and twentieth-century agrarian life. The historical records are limited in scope, and to a degree the poverty of the sources shapes the direction of the inquiry. In this chapter I use these societies' encounter with maize as a means to examine the changing pattern of daily life in these agrarian empires.

The sixteenth and seventeenth centuries brought changes in the economic and intellectual domains in the Mediterranean world that required responses from both peasant farmers and elite classes. The responses to maize of the cultural elite emerge in historical sources as aesthetic, pragmatic, and sublime. But the symbols of the New World expressed themselves quite differently in the world of the cultivator, the part-time merchant, and the local cleric. The miller's gate, the brewer's hut, and the bustling regional marketplace were all locations for the exchange of ideas, places that the sixteenth-century satirical poet Pietro Neli describes as "the soft ground muddied by the piss of the village mules." This earthy and evocative image pertains as much to the ambiance of fieldwork interviews and participant observation of peasant culture in late twentieth-century rural Ethiopia as it does to life in sixteenth-century Italy, as we can reconstruct it from the archival traces.

In such muddied places peasant intellectuals like Carlo Ginzburg's irascible miller Mennacchio "must have talked about many things."[2] Though the paucity of historical sources leaves much to our imagination, these "things" must have included the exchange of news between neighbors and kinsmen about crops, as well as novel ideas about the scope of God's universe, the properties of unfamiliar seeds, and the excessive intervention of landlords into farmers' business. We may need to remind ourselves that Galileo Galilei's lectures on the nature of the solar system and astronomy from his university lectern in Padua paralleled the appearance of maize in the fields that surrounded his scientific studio. The arrival of maize on the scene was, arguably, at least as earth-shattering an event in local lives as his blasphemous view of the universe in those same years.

It seems likely that farmers in Italy and in Ethiopia alike may have been curious about the new food crop that had percolated into their universe—and may have experimented with it out of sight of their elite overlords. The results were in one case (the Veneto) almost immediate adoption and in the other case (highland Ethiopia) a three-hundred-year delay in full acceptance, but no less radical a final outcome.

Maize arrived in both northeast Italy and Ethiopia shortly after 1500, as part of the massive global ecological and demographic transformation of the Columbian exchange.[3] As is true for West Africa, however, little documentary evidence records what must have been a conscious process of Europeans and Africans' introducing this botanical curiosity into elite gardens or to peasant farmers' fields. Those who advocate peasant agency and those who favor the idea of patrician innovation will find equally little to document their assertions.

Luckily, we can sketch an approximate chronology for the first incorporation of maize seeds into particular Old World farming systems, describe the nature of the farms' political ecology, and offer educated guesses about the genetic personality of the maize types that first presented themselves to farmers as an agronomic option. As to type, the first maize seeds that reached Old World farmers seem to have been Caribbean and Brazilian flint maizes (in a reflection, not surprisingly, of Iberian trade links). Characterized by their hard starches, relatively longer storage capacity, and early maturity, these distinctive maize types fit into special niches in Old World agriculture. Like most maize types, flint maizes also varied in the color of the kernels, the height of the plant, and the shape of its husks. Flint maizes generally were hardier than other types and were the first to adapt to colder North American sites, such as New England and the upper Midwest, or to the semitemperate settings of highland Ethiopia and northeast Italy. Flint maize's hard starch and early maturity suited it well for feeding hungry farm families and for making the stiff porridge—polenta—that would become Veneto's plebian staple.

The Venetian Agrarian World

In the year 1500 the Venetian geographer Amerigo Vespucci used the term *mundus novus* (New World) in a treatise describing the new lands of the Western Hemisphere as reported on by the Genovese sailor Christoforo Colombo. Vespucci's intended reference was not merely to geography: he sought to invoke abundance, the sublime, and endless possibilities. To the more perceptive among Venice's patrician and mercantile classes the term had a special ironic ring. As perhaps the world's most globally aware merchants, Venetians must also have been the first to recognize that the opening of the Atlantic and Indian Oceans to European commerce sounded the death knell of their own knowledge-based prosperity.[4]

Over the period stretching from the sixteenth to the eighteenth century, Venice progressively lost its international commercial grip, owing to the decline of the maritime grain trade and the atrophy of its eastern trade, as its hegemony in those areas was usurped by the Dutch and the Portuguese, respectively. In the face of an ebbing of its mercantile monopoly in the eastern Mediterranean, an increasingly unprofitable industrial and artisanal economy, and a collapse in grain prices, Venice turned inward to its own *terraferma* in the Veneto, Friuli, and Lombardy as the site for conservative investment in landed estates. There Venetians rediscovered a peasant economy they could exploit as a tax base and as a guarantor of Venice's food supply.[5]

The agriculture of northern Italy in the sixteenth century, however, contained within it an essential disequilibrium. Its property system pitted estate owners' interests in controlling land holdings and the right to income against peasants' struggles for security of daily subsistence and land tenure. Moreover, the wheat, the region's cereal staple and consumer crop of choice, was unreliable both as a food and as a taxable commodity.

The farms of northeastern Italy were to be the locus of significant agronomic change from the sixteenth through the eighteenth centuries. The agent of profound transformation in northeastern Italy during this period was *Zea mays,* a plant that appeared initially here and there in Venetian gardens and herbaria, and whose image began to appear in frescoes and sculptures in hunting lodges and ducal palaces. Maize began its life in Italy first as a curiosity sown a plant at a time in household gardens in Venice and its rural hinterlands, but by the eighteenth century it had become a staple crop overwhelming all other foods and occupying the lion's share of peasant acreage in Lombardy and the Veneto.

Venetian food ways underwent a remarkable metamorphosis when maize arrived. Before 1500 the agriculture of northeast Italy stood foursquare within the European regime of winter wheat supplemented by *biade minute,* or minor cereal crops, such as rye, sorghum, and millet, that fit, albeit awkwardly, into crop rotations with winter wheat and local vegetable and tree-borne crops. Wheat was king among Europe's prestige food crops, and the bread made from northern Europe's soft winter wheat counted as the symbol ("daily bread") as well as the substance (the primary foodstuff in the region) of well-being. In most of Europe soft bread wheat was the primary grain, though by the seventeenth century the bread of Italian towns was often blackened with imported Baltic rye when wheat crops failed. For the peasants of northern Italy wheat was also a symbol of elite authority: wheat served as the required medium for in-kind tax remittances to the state and rent payments to the landlord.[6]

The tyranny wheat exercised as the principal grain depended on its precarious agronomy. Planted in October and November and harvested in late spring and summer, winter wheat was highly sensitive to fluctuations in temperature and moisture. The minor cereal crops—the *biade minute*—never challenged wheat's dominant role either agronomically or symbolically. As a cold-climate cereal crop, rye was an alternative to wheat and agronomically safer, but it had too low a yield and fetched too unrewarding a price to attract Veneto farmers or their landlords to cultivate it extensively. Summer grains of African origin like sorghum and millet required long growing seasons and could not reach maturity within the few months between winter wheat's spring harvest and its fall sowing, and thus could not easily fit into a regular crop rotation pattern. Sorghum and millet also had exposed grain "panicles" and thus were highly vulnerable to bird damage; anyway, millet served primarily as livestock forage, rather than as human food. Moreover, on the region's low lying plains the spring flooding in Italy's Po Valley restricted wheat to the limited tracts of cropland immune to the spring inundation. In fact, some of the seasonally flooded alluvial soils had already shifted over to northern Italy's distinctive but geographically limited rice cultivation.[7]

Agriculture in northeast Italy—especially in the Po River valley—was a historically dynamic system linked to both urban markets and estate management but constrained by what agricultural economists would call a fundamental bottleneck in crop-livestock integration. Most seasonally flooded land provided only a single harvest, after which the plot served as summer pasture and a limited source of hay for winter stall-feeding of cattle.[8] Because winter wheat occupied land that might otherwise have served as winter pasture for livestock (both draft animals and dairy herds), farmers had to transfer livestock to winter pasture in pre-Alpine foothills.

Thus, Po Valley farms annually lost a quarter of their precious manure potential. Farms in the plains of Lombardy, the coastal floodplains of the Veneto, and upland zones around Alto Adige all had to deal with the vicious cycle of poor feed for livestock, the shortage of manure to maintain soil fertility, the need to produce marketable cereal crops (especially wheat) for the state, and the annual struggle to maintain an adequate rural food supply. At the same time, annual crop production had to provide the food and economic sustenance needed to support investment in perennial nonstaple crops such as wine grapes, mulberry leaves (for silk production), flax, hemp, olives (in forested areas around Lake Garda), and fruit trees that helped sustain upland area economies.

At the center of this production paradox was a simultaneous need for human food and forage for livestock. Breaking this cycle required expanding livestock's food supply and making use of seasonally flooded lands that could not support wheat production. Harnessing water and transforming lowland wetlands from seasonal floodplains into arable fields was one key to the cultivation of both winter wheat and summer crops. As early as the twelfth century Cistercian monks had begun systematic harvesting and management of water on a small scale in the western Po Valley south and east of Milan. By the sixteenth century large portions of the low plain were irrigated, and seasonally flooded lands were under flood control schemes designed by civil engineers and financed by city-based investors. This technological change and economic investment combined urban capital, rural labor, and specialized engineering expertise to transform the relationship between nature and human economy.[9] The use of urban commercial capital to create, in effect, new cropland under direct control of the urban investors increased landlords' leverage over peasants who depended on sharecropping contracts for their access to land.

Over the course of the sixteenth through nineteenth centuries an increasing percentage of such reclaimed land, moving into the regional agricultural economy, converted it into an integrated agricultural system of annual cereals (wheat, rice), livestock, perennial vineyards, and tree crops. The early practice of stall-feeding of livestock in Lombardy and the Veneto also fostered local markets for livestock forage. Simultaneously, demographic pressure pushed agriculture onto marginal lands that required early-maturing crops (to avoid destruction by spring floods) and created an expanding demand for fertilizer (manure from domestic livestock).[10] Circumstantial evidence suggests that peasants surmised much more quickly than complacent conservative estate holders that the adoption of maize would cut this Gordian knot and offer some radical alternatives to the structure of the agrarian economy.

The opening of the *mundus novus* constituted something of a spiritual and economic apotheosis for the northern Mediterranean and its heart, Venice. The news of the New World had shown literate Venetians the handwriting on the wall, and Venetian capital sought a safer haven than commerce. Patrician families in Venice thus began to launch new investments in rural holdings on the Italian mainland that seemed safer than mercantile voyages. Those investments flowed into the water management and rural estates mentioned above that extended Venetian control over rural lands and opened new year-round production.

Italy's Po Valley occupies a plain that extends north and south of 45 degrees north latitude—that is, the same latitude as the U.S. corn belt. Moisture—the usual "limiting factor" for maize cultivation—is not an issue in the Po Valley because the presence of the Alps and Piedmont to the north creates high humidity as well as a source of water for riverine and canal irrigation. The climate of both the plains and the hills of the regions of the *terraferma* falls well within the temperature and humidity range ideal for maize.

In his compendium on maize in Italy, *Il mais e la vita rurale Italiana,* Luigi Messadaglia states that a certain Venetian diplomat named Andrea Navagero visited a Venetian botanist, Giovanni Ramusio, in Seville in the 1520s and saw maize being cultivated. On his return to Venice, Navagero carried some maize seeds, most

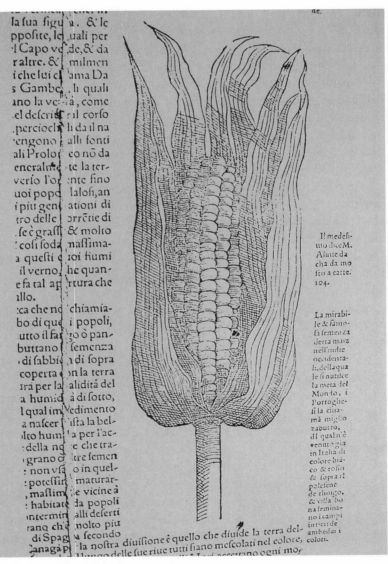

9. Maize: drawing by Giovanni Ramusio, ca. 1540.

probably some of the Cuban Caribbean red flint that Columbus had first brought to Seville after his second voyage. By 1554 farmers in Polesine (southwest of Venice) had their first fields of maize.[11]

Maize formed part of a new economic beginning that had representations in the aesthetic world of the late Renaissance even before it appeared in its gardens and on its farm tables. As early as 1515, in Giovanni della Robbia's terra-cotta sculpture *Tentazione di Adamo* depicting Adam's fall, the scene contains images of maize plants within an exotic Eden. At the geographic and spiritual heart of Venice, the Palazzo Ducale near the Piazza San Marco, the marble decorations on the canal portal constructed around 1550 (directly below the Bridge of Sighs) contain several ears of maize sculpted among the familiar fruits and vegetables of the cornucopia. While maize began to appear in botanical lists and in gardens around Europe by the mid-fifteenth century, it took on names like *Turcum frumentum* (Turkish wheat) and *Frumentum asiaticum* (Asian wheat) indicating a general European confusion about its origins in the New World.[12]

Iconic images of maize appeared in domestic settings as well as in public art. Among the bucolic frescoes (painted around 1540) decorating the walls of the hunting lodge of the Emo Capodilista family in the Euganean Hills, in the rural hinterland of Venice, are several painted depictions of ears of green maize among images of fruits and vegetables decorating an internal doorway.[13] The early presence of maize as an iconographic symbol of abundance and exoticism suggests an expanding worldview within the Venetian consciousness at the same time that it asserts the elites' myopic view of maize as an exotic vegetable, rather than the dietary staple cereal it would become.

A more profound indication of the cognitive acceptance of maize within northeast Italy is the variety and subtlety of regional and dialectical names used historically for the plant and its food products. Venetian names reflected themes similar to those of their African counterparts in choosing to designate the stranger by familiar

10.
Maize:
fresco
in Colli
Eugenai,
Veneto,
ca. 1540.

11. Maize carving on the door of the Palazzo Ducale, Venice, ca. 1550.

descriptive terms. We can read these names to gain insight into peasant perceptions of maize, which stood in direct contrast to the esoteric elite views of maize featured in representational art. As in Africa, the names ascribed to maize reflect some combination of its resemblance to known crops, especially sorghum *(sorgo)* or wheat *(frumento* or *grano)*, other aspects of its appearance *(rosso, zalla*—red, yellow), and its perceived provenance *(turco, moriscu, Indie, moro, Sicilia, saraceno)*. One agricultural dictionary cited by Messedaglia offers a motley lexicon of Italian regional and dialectical names for maize: *formentone, frumentone, granoturco, sorgo turco, grano d'India, grano siciliano, grano di Roma, melgone, melicone melica, meliga, miglio zaburro, formentazzo, melgotto, melicotto, carlone, madonnino, polenta.* He also lists the words for maize in seventy-four regional dialects, including the Venetian dialect's *zalla* (that is, *giallo*, yellow); *soturco* (from *sorgo turco*, Turkish sorghum); and *strepolin*.[14] The generic terms for maize in standard modern Italian, *granoturco* and *mais*, predominate today, though local dialectal terms for different varieties still occur quite commonly.

Along with the historical insights derived from examining the naming of maize by its peasant hosts, considerable historical value is to be gained from understanding the personality and characteristics of the maize itself. What arrived in Veneto and the Po Valley in the early sixteenth century was a maize type that reflected the ecological history of New World maize and the nature of the trade network that linked the eastern Caribbean with the Mediterranean— namely, a Caribbean red-flint maize type descended from the seeds brought by Columbus to Seville. In Italy the classic flint corns were the ancestors of the maize used still for polenta, which predated the modern hybrid dent varieties that arrived in the late 1940s and like most maize globally serve now primarily as livestock feed.

Northeast Italy's historical varieties of maize were all flints distinguished by their early maturity and hardiness. *Cinquantino* was the most common descendant of the Caribbean flint imports, so named because of its fifty-day maturation.[15] *Quarantino* was another classic flint variety, which matured, reportedly, in only forty days. *Marano* (or *maranino,* a version of *cinquantino*) had distinctive red kernels that produced an orange-tinged polenta, the staple of many generations of Veneto *contadini,* and may have been the type of maize that peasants originally dubbed *sorgo rosso.*[16]

Though only obliquely appearing in historical records, the transformation of the cropping system and food ways of peasant farmers was an important struggle played out within the agrarian political ecology of northeast Italy. Extant tenant contracts in the Venetian records make it clear that landlords maintained extensive documentation on each of their holdings, including the exact surface area, dimensions, soil type, and anticipated yield. Michele Fassina describes an explicit contract in Vicenza that detailed exactly the condition of the fields at the beginning of the five-year contract, the crops to be sown in each year, and the condition the fields were to be left in at the end of the contractual term.[17] As

12. Red flint maize, Ethiopia, 2002.

property owners, landlords had the legal right to alienate land and terminate a tenant's use of the land. Tenant contracts specified in formal terms precisely the condition of the fields at the beginning of the contract, the prescribed crop rotation to be followed, the areas to be left for pasture, and terms of rent collection.[18]

The preference for rent payment in wheat and direct control over cultivation and management was a source of both subtle and overt struggle between cultivators and landlords. Venetian property law allowed landlords to set the terms of peasant access, while population growth over the sixteenth and seventeenth centuries generally increased landlords' leverage over their cultivators. These terms favored wheat at the expense of peasants' security of tenure and their household food supply. For their part, tenants struggled to assert their own moral economy of entitlement to land as familial property that was heritable and partible and as a resource to guarantee subsistence.

Within the first years after the appearance of maize on peasant landholdings, landlords apparently allowed peasants to cultivate maize as a *biada minuta* (secondary crop), a category to which landlords paid little heed, as it constituted their tenants' food supply and did not affect their rent payments in wheat. All over the region, however, it became apparent in the late sixteenth and early seventeenth century that maize's impressive yield and perfect fit in the summer crop rotation mix had revolutionized both the tenants' food supply and their cultivation habits, in the process supplanting rye, sorghum, and millet almost entirely. Maize allowed peasant farmers to put into summer cultivation plots that flooded in spring and had heretofore been simply seasonal pasture or agriculturally idle wetlands. Moreover, it also became clear that maize served a dual role as human food and as silage and stover, which filled the gap in livestock forage and allowed winter stall-feeding of cattle, and as a consequence expanded manure supplies and on-farm dairy production.[19] Maize as a fodder crop also complemented the expansion of water control in the Veneto by converting what had been seasonally flooded pasture into cropland cultivated year round in a wheat-maize rotation.

The early entry of maize into peasant plot rotations and into tenant-landlord disputes over plantings mostly falls outside of the purview of the written historical record. Nonetheless, a sketchy chronology is possible. The first official estate record of maize is a 1582 *polizza de estimo* (estate sowing plan) from landlord Pier Maria Contarini that indicates that his tenants planted maize *(formenton zalo)* on a seasonally waterlogged plot on his family's estate at Vighizzolo Vinico Este. But maize cultivation must have been expanding informally on tenant holdings for several years prior to that, because in 1584 maize appears in the postmortem farm inventory of a peasant in a Vicenza village, and four years later maize made up part of a gift from a peasant farmer to the monastery of St. Bartholomew of Rovigo. By 1601 an official document from Venice's Rialto market states that maize was "bought by the most impoverished and miserable persons, for whom it is a source of sustenance with a good market."[20] During this initial expansion maize crossed the threshold from being a subsistence food to become a market crop with local and then regional effects. Off-farm sales suggest increasing urban consumption. Peasant smallholders accomplished this transition by expanding their own plantings and exploring the various ways they could plant the crop.

Maize had begun to increase its public visibility in the early seventeenth century. In 1604 Galileo published an imaginary dialogue between two peasants in Paduan dialect to explain his controversial new theories about the heavens and a nova that had appeared suddenly in that year among the visible stars and planets. In his dialogue one peasant suggested a name for the nova to a disbelieving companion: "Call the new star 'quintessence,' or call it polenta!" It could be observed nonetheless.[21] Maize was evidently already part of his personal universe and perhaps his daily diet.

Historian Michele Fassina argues that by the early seventeenth century maize had become a staple that appeared in every province of the Venetian republic. In 1617 the major estate of the influential Contarini family at its estate at Piazzola sul Brenta for the first time accepted rent payments for land in maize. In another confluence of the new scientific order and realities on the farm, two years after the Papal Edict of 1616 by Pope Pius V that forbade Galileo to speak of a heliocentric cosmos, the Venetian state finally accepted maize as payment for tax levies. Heaven and earth in northern Italy were never quite the same after these coincidental events. With its 1618 decision to accept tax payments in maize, the state itself recognized the agrarian constituency's insistence on the crop; by officially accepting maize for the payment of taxes, Venice symbolically signaled the end of what must have been a much longer agrarian conflict. And during the 1630 Venetian plague, maize made up a large portion of the food sent by the state into that quarantined city haunted by the carnival images of death.[22] Fassina's study of Polesine farm records from a single holding during the years 1718 to 1764 indicates that maize by that time had become firmly established as the major crop in the region, accounting for more than 50 percent of all farm income, and in some years as much as 70 percent of the total.[23]

Thus, the initiative to adopt maize derived from farm-level decisions by a peasant population chafing under restrictive property law and tenuous access to land. The chronology of maize's appearance in official records and recognition by both landlords and the state was largely a response forced by peasant initiative and the crop's success on the farm itself. This initiative resulted partially from the weakness of the cropping system that perpetuated the privileged elite's preference for wheat, a crop poorly suited to the agroecological setting of the Veneto. But it also reflected peasants' choice of maize production as an avenue of resistance to combat both their lack of access to land and the efforts of the elite to maintain control over farm management. The adoption of maize opened new farm-level options in crop-livestock integration, in addition to establishing new terms of sumptuary class divisions and rural-urban differentiation. Maize was thus both the symbol and the substance of peasant response to their plight.

Peasant monocropping and the advent of maize as the dominant food source would reveal their most serious consequences only in the nineteenth century, with the emergence of endemic pellagra among rural maize eaters, an affliction that came to symbolize their poverty.

Maize and Ethiopia's Solomonic Empire

The Venetian elites' appropriation of maize as a symbol of the New World's natural abundance and their own worldly opulence stands in sharp contrast to the silence of the historical record on the arrival and reception of maize in Ethiopia. *Zea mays*'s first appearance in the historical record is Caspar Bauchin's supposed observation of it (1623).[24] But where and how did it arrive there? It is often argued that Portuguese missionaries brought seeds with them in the early years of the seventeenth century, presumably as part of their program to broaden the new brotherhood's ecclesiastical ties to the Land of Prester John and their successful attempt to convert the then Ethiopian emperor Susenyos to Roman Catholicism, an event that took place in secret in 1612.

The oft-repeated story goes that the Jesuits brought maize seeds with them from their mission outpost at Goa (off the coast of the Indian subcontinent) or directly from Iberia. Maize would have been for them a tropical food crop that they imagined would bring innovation and sustenance to a new community of believers and their mission base. After all, initial Jesuit successes in Ethiopia (Abyssinia) had included receiving a land grant at Fremona in Tigray from the grateful emperor. Indeed, there is circumstantial evidence that suggests that the maize that reached Ethiopia in those early years was probably the same red Caribbean flint maize that had moved from Spain to northeast Italy and thence into the Mediterranean and Near East in the mid-to-late sixteenth century.

Unlike the aesthetic evidence in the Venetian case, however, there is no reference to maize in any of the Ethiopian royal chronicles or in any iconography from the period, despite the significant evidence of the Portuguese (and other European) success in influencing architectural style, iconography; the conversion of Emperor Susenyos and the imperial court to Roman Catholicism was part of that process. Indeed, the anti-Jesuit counter-revolution of 1632 rejected symbolically and with violence precisely these foreign influences associated with the Jesuits and externally generated modernity. Moreover, the otherwise voluble Portuguese accounts of their activities in Abyssinia make no claim to introducing new crops as part of their mission.

The argument that it might have been the Capuchin missionaries who arrived in Abyssinia in the early seventeenth century who introduced maize to the northern highlands is weak as well. The visitor (Caspar Bauchin) who wrote of seeing a kind of pod maize there in 1623, was reporting an observation and not a claim to have introduced it himself. Moreover, the Capuchins arrived in Ethiopia via the Nile Valley, and the ubiquitous name for maize in northern Ethiopia, *yabahar mashila* or *mashila beheri* (sorghum from the sea), implies a maritime provenance, such as the Red Sea or Indian Ocean.

While historical evidence from the seventeenth and eighteenth centuries relating to agriculture is relatively scarce, there is no clear indication from the Ethiopian highlands of the kind of agrarian crop revolution that took place in the Veneto. The agriculture that reemerges into visibility in the early nineteenth century appears to be the same diversified multicrop farming system described in detail by the first Portuguese visitors three hundred years before. But maize does appear in some accounts, as a spring crop in gardens and a few fields, more a snack than a staple. Clearly, maize had not overwhelmed Ethiopian peasant agronomy as it did so thoroughly in the Veneto.

Was the peasants' seeming rejection of maize as a field crop linked to anti-Catholic xenophobia? The idea has some merit. After all, the Amharic and Tigrinya terms for maize (the sea's sorghum) referred directly to its foreign provenance. The Ethiopian emperor's abortive conversion and the quick and successful counter-reformation led by Susenyos's son Fasiladas may in fact have been broadly reflective of a regional intellectual and social conflict over rationalism and pragmatism in religious and commercial life in the Nile Valley in the seventeenth century.[25] It is difficult to say, but the argument for a Portuguese introduction of the grain is hard to sustain in view of the total lack of evidence.

The most plausible hypothesis for arrival of maize in Ethiopia comes from Merid Wolde Aregay, who is deeply versed in the historical literature of the seventeenth century.[26] He argues that we ought to take the silence of the Portuguese sources as evidence of their noninvolvement. Rather, he states, the pattern for other items of trade and exotica, such as Indian cloth and spices, was one of introduction via Arab and Banyan merchants whose commercial networks linked Ethiopia's Red Sea ports to the Persian Gulf and Indian Ocean trade. Thus the same agents that brought maize (*muhindi* in Kiswahili) to the Swahili coast may have also carried it to the Horn of Africa. The diffusion of maize into the highland agrarian economy would have resulted from slow percolation, rather than a transfiguration radiating outward from the royal court.

Unfortunately, the historical records of the

13. Pod maize, South
Africa, ca. 1914.

period from Ethiopia offer no insights into whether maize had a value to the elites as either a symbol or a food source. Equally, there is little or no evidence that Ethiopian highland farmers treated maize as anything more than a kitchen garden vegetable and agronomic curiosity; its adoption as a field crop would take several centuries. However, the agronomic exigencies behind this long delay in adoption are worth examining.

Ethiopia's highlands have historically displayed a remarkable stability both in technology during the evolution of a farming system and in the constitution of a repertoire of food crops. Deep soils, dedicated farmers, and a distinctive African technology— the single-tine scratch plow called the *maresha*—promoted early innovation in cereal crops. Moreover, the agronomic strategies deployed in the highland's kaleidoscopic microecologies, created by the highlands' array of elevations, soils, and rainfall shadows, made highland agriculture a complex and resilient enterprise. By 1500 the plow, along with a classic bundle of endemic cereal cultivars (barley, teff, wheat, sorghum, finger millet, and so on) and an annual cropping regime, had existed for perhaps two millennia or more amid the profusion of local polities, ethnicities, and ecological frontiers.

Highland Ethiopian agriculture was finely tuned to the "limiting" factor of elevation and its effects on both temperature and rainfall. Crops, rotation systems, and seasonal rhythms of planting, harvest, and seedbed preparation fit a bimodal climate regime that had evolved over many generations. Ambilineal, partible patterns of property inheritance and the exigencies of microecologies of soil and elevation meant that farmers generally controlled a series of small plots spread across several agroecological zones, each planted according to its own rotation scheme. Ethiopian women did not plow and had little control over plot management, but they nevertheless enjoyed direct influence over the processing of food and the garden crops *(gwaro)* grown in the fertile soils around the house itself. There herbs, vegetables, and condiment crops complemented the grains, beans, lentils, and herbs of the farms' main plots. Thus, maize may have gotten its start in Ethiopia as a woman's plant, cultivated horticulturally rather than as a field crop.

What accounts for peasants' conservatism in northern Ethiopia, as compared to Italy? In contrast to northeast Italy, in Ethiopia crop choices—whether to plant teff or wheat, sorghum or finger millet—reflected a finely tuned judgment and were controlled only by the farmer himself. The farmers' adage "Maret ende mert" ("Do as the plot chooses") indicates that crop choice was decidedly a local matter. The saying also hints at not only the ecological nuances affecting local plot conditions, but also the historical fact that elites who held the bundle of income rights that included a tithe and rights to corvée labor from rural cultivators had no such rights over farm-level management, choice of crops, or the distribution of land among those with right of usufruct *(rist)*.[27] This flexibility stands in opposition to the strict contractual terms imposed by Venetian landlords on their *contadini*. In Ethiopia's Christian kingdom, by contrast, few sumptuary laws existed, and status was less often used as a gauge to separate rural social classes; the facts of poverty nevertheless meant that members of the elite ate more of the prestigious grain teff and more protein than did those from whom they claimed labor and payment in rural products.[28]

At the beginning of the seventeenth century, the political ecology of agriculture in highland Ethiopia was highly elaborated in both its agronomic and its political practice. That elaboration resulted in rural producers' adherence to conservatism in both religious practice and property rights. Elite rights of collection *(gult)* held implicit within them a recognition of peasant claims on heritable rights to land use *(rist)*. These entitlements were embedded within a system that also accepted a set of vertical power relations in which the political and ecclesiastical elite occupied the rung.[29] Smallholder rights to farm-level management and claims to land use cemented a conservative moral economy among farmers with regard to changes in the hierarchy of authority, an attitude that resulted in farmers' strong sense of entitlement over access to land and crop management. Ethiopian farmers already had the crops suited to local microecologies and, in contrast to farmers in the Veneto, few political incentives to change.

The circumstantial evidence available suggests that in highland Ethiopia maize had little effect on diet or farming. Rather, maize remained nestled in the farming system as a garden vegetable crop probably cultivated by women and consumed green before the major cereal harvest, or as a field crop that occupied a narrow ecological niche (determined by altitude) between wheat and sorghum. Maize remained a minor crop because peasants had little need of it and the elites had no means or desire to insist upon its cultivation.

Like their counterparts elsewhere in Africa, Ethiopian farmers who encountered maize appear to have added it unceremoniously to their complex crop repertoire.[30] Maize first appeared as a grain in historical records of the early nineteenth century. In 1805 traveler Henry Salt noticed in Tigray a well-established field of *e bahr mashella* (the sea's sorghum), a term he glossed as "Indian corn." He described a valley "well cropped, especially with Indian corn, which is usually more forward in this climate than any other grain," and he also identified maize (probably a flint type) by its most salient trait, early maturation.[31] Other nineteenth-century travelers also found it here and there throughout the highlands, but never as the dominant grain crop it had become in the Veneto more than a century earlier. Most of these descriptions make it clear that what visitors saw in Ethiopia was a yellow flint maize with a low yield, a plant height "that does not exceed 1.25 meters and has ears of ten rows of small grain."[32] One could speculate that, as a flint maize, it was favored by women not because of any advantage in yield over other grains but because of its early maturity and suitability for milling on a grinding stone or with a mortar and pestle. Many consumers also liked the fact that it provided fresh food long before any of the field cereal crops were ripe and edible, a factor

14. Local Ethiopian maize varieties.

Note: Clockwise from the top are a crimson flint; a blue, white, and crimson flint; a blue and white mixed flint; a yellow flint with white highlights and odd kernels of blue; a uniform red flint; a reddish semident with white highlights; a rosy flint; a blue and white flint; a white dent; a red and purple dent; a yellow dent with a single purple kernel; a red and purple semident; a blue and purple flint; a mixed semiflint with blue kernels, two rows of rust and white highlights (Twumasi's favorite); a blue and white flint; a rose semident with a few purple kernels; and a brownish orange flint.

that appears repeatedly across the continent. Since farmers more often planted maize in the women's household garden, as a vegetable, than in the men's fields, it was sown on ground opened by the hoe rather than the plow. Maize never became dominant in the plots devoted to it or in the peasant diet, and Ethiopian farmers' dedication to a diverse range of locally adapted crops continued into the modern era.

Apart from the red herring about its Portuguese origins, however, another tantalizing suggestion has surfaced regarding the crop's introduction to the Ethiopian region. A much later avenue is suggested, leading from the southwest, perhaps via East Africa's Great Lakes region. The best evidence of multiple points and times of entry is the enormous variety of maize types and colors still evident in local maize varieties in those regions of the country. They include red flints and semiflints, red dents, purple, mixed yellow and white, and variegated colors.

The impact of maize on the Ethiopian region as a whole comes clear in tantalizing bits on examination of the geographic distribution and linguistics of the names given to maize by thirty-five of Ethiopia's eighty-five distinct languages. Discernible clusters of names for maize conform both to language families (Semitic, Cushitic, Omotic, and Nilo-Saharan) and to key borrowings across language families. The patterns reflect not only naming trends elsewhere in Africa (for example, "sorghum" and "from the sea") but also distinctive names for the stalk, ear, and leaf. Overall, however, the Semitic for "sorghum from the sea" and the Cushitic terms *mashila* and *boqolo* used by Semitic and Cushitic speakers alike in Ethiopia suggest at least two avenues of contact early on. One was from the north, and probably based on Red Sea trade contacts in the late sixteenth or early seventeenth centuries. Languages and regions that chose *yabahar mashila* (sorghum from the sea) or a shortened version, *mashila,* in Gojjam reflect that point of arrival. The analogy to sorghum also strongly suggests that the groups in the northern highlands that favored cereal crops and use of the plow perceived the new crop as a grain akin to sorghum, rather than as a vegetable.

The widespread use of the term *boqolo* or *badalla* to refer to maize in the Cushitic and Omotic language families, however, suggests another point of penetration from the southwest, probably via the Nile Valley, the lakes region, or both, perhaps somewhat later. *Boqolo* appears as the predominant term in some Omotic languages (Walayta, Shinasha, Gamo) and some Cushitic languages (Oromifa, Hadiya, and Alaba), but the term *badalla* (Walayta, Sidama, Koorete) seems confined to Omotic, except in Gedeo, a Cushitic enclave surrounded by Omotic languages. The geography of maize word groups and possibly the patterns of influence, where horticultural systems adopted maize earlier in the southwest as a vegetable garden crop, were precisely those areas where northern highland cereal-based systems had had the least influence. Here the coining or adoption of the term *boqolo* may reflect the local ideas of social systems in groups whose focus was horticultural, in contrast to more northern societies that practiced cereal-based farming. The adoption of the term *boqolo* in most Amharic-speaking areas and in virtually all Cushitic areas may indicate the strong Oromo influences of the sixteenth and seventeenth centuries; in that case, it would be possible to restrict the *yabahar mashila* zones to those areas where Oromo economic and cultural influences were less systematically felt. Maize would have been a very late arrival indeed in the Eastern Cushitic–speaking areas of Somali and Afar speakers where sorghum's drought-resistant traits would have caused farmers to see it overwhelmingly as the crop of choice. In those areas, and in the eastern Harari enclave, the use of the term *Arabih* (Arab sorghum), reflects an orientation toward grain crops in the pastoral context, and the trade links to the Red Sea or Indian Ocean that characterize northern Ethiopia.

Ultimately, maize's role as a dominant field crop began along the southern periphery of the Ethiopian highland kingdom, among Cushitic- and Omotic-speaking farmers rather than on the classic cereal-based farms of the northern highlands. In southern regions maize had found a rather warmer reception within hoe-based and plow-based farming systems that cultivated it as a vegetable crop to complement the false banana *(ensete ventricosum)* plant as well as other highland crops. In forested zones, maize served, as it had in West Africa, as a pioneer crop in land cleared of forest.[33] It seems probable that these southern agricultural systems were part of a quite separate maize complex where the crop arrived from the south, via the Nile Valley and Nilotic culture areas to the south and west that had adopted maize in the mid-to-late nineteenth century. Names for maize in southern and western Ethiopia have Cushitic roots *(boqolo)* rather than the Semitic *(yabahar mashila)* connection. Recent collections of "farmer varieties" of maize in southern Ethiopia show great richness in types and colors, including classic red flints, yellow dents, and the *cinquantino* flint identified as early as the 1880s.[34]

In 1938 an Italian colonial survey team led by crop scientists Rafaelle Ciferri and Enrico Bartolozzi compiled statistical tables on cereal crops (including maize) cultivated in Italian East Africa—an area including Eritrea, Ethiopia, and Somalia—which indicated the distribution of cereal crops across the six colonial Italian regions. Maize was a major crop in the south (Galla and Sidama), but in 1938 it had only a minor presence in other areas. Overall, maize accounted for just 14.5 percent of Italian East Africa's total cereal production, ranking well behind sorghum, barley, and teff as a primary staple.[35]

The Italian colonial data on maize from 1938 are less an indication of the colonial influence on Ethiopian agriculture than a snapshot of Ethiopia's traditional cropping systems at the dawn of the modern agricultural era in that economy. While the Italian colonial figures offer somewhat imprecise estimates based on a broad survey, they nonetheless give a fair approximation of the situation in the agroecologies of the Horn of Africa in the second third of the twentieth century, for in many ways the regional boundaries Italy imposed approximated both cultural and agroecological zones. The central highland agricultural zone tilled by classic ox plow, for example, that made up the Amara region displayed a balance between the highlands' traditional cereals, teff and barley, and the sorghum that dominated the lowland periphery. The relatively mi-

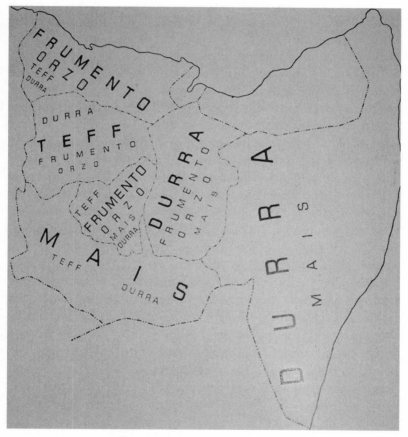

15. Cereals in Italian East Africa, 1938.
Durra: sorghum; Frumento: wheat; Mais: maize/corn; Orzo: barley;
Teff: teff *(Eragrostis teff)*

nor penetration of maize into Amara, Eritrea, and the Shawan regions (essentially the highland ox plow farming system) indicated that its long-established role as a minor garden crop remained largely unchanged.

In all regions except the southern zone of Galla and Sidamo, maize was a minor crop. In Galla and Sidamo, the region that included the south and southwest of Ethiopia, however, maize was the dominant crop, surpassing all others. Its overall predominance in that region, however, masked the fact that it appeared in a wide variety of farming systems, including horticultural intercropping with ensete, garden cropping in highland teff zones, and cultivation as a field crop complementing livestock in better watered areas. Southern farming systems depended heavily on maize as a cereal staple, but agronomically maize also appeared on their farms and in their food ways as an intercropped vegetable with root crops, such as ensete, cassava, and the local root vegetable *Coleus edulis*.

During the Italian occupation, at least one episode occurred that had maize as its subtext. During the course of its short-lived colonial enterprise in Ethiopia (1936–1941) the Italian Fascist government attempted to address Italy's chronic wheat shortage by launching a program of *panefìcazione* ("breadification") in its new East African colony. This program created local immigration schemes that brought entire Italian peasant families into settlements in key agricultural sites in the Ethiopian highlands for the purpose of cultivating wheat for the Italian market. Ironically, these recalcitrant Italian colonial peasants imitated their Venetian forbears, and after only a year of half-hearted effort the transplanted *contadini* abandoned wheat and sowed maize, much to the consternation of their *Fascista* overseers.[36]

By the second half of the twentieth century, a new export economy for coffee in southwestern Ethiopia created a conflict over labor, land use, and the agricultural cycle of traditional annual cereal crops. Instead of collecting wild coffee from the forest—the nineteenth-century method—farmers began to propagate coffee seedlings and to integrate coffee cultivation into farm production. Maize, already cultivated here and there in southern cropping systems by the late nineteenth century, proved an ideal complement to coffee in the demands it made on farm labor. In the period after World War II there evolved a widespread coffee-maize complex that paired coffee as a cash crop and maize as an early-maturing, low-labor food crop that provided sustenance before the coffee income came in, in the fall. Those farmers who did not produce coffee benefited from off-farm cash income as coffee laborers and could still sow maize on their own plots. By the 1950s this coffee-maize cropping system was well entrenched and increasingly included improved dent maize introduced by the early agricultural extension programs sponsored by foreign donors, especially the U.S. Point Four (later USAID) program, via Ethiopia's new agricultural colleges and research institutes.[37]

The coffee-maize complex in southern Ethiopia, in many places still tying into the culture of *Ensete ventricosum* ("false banana" plant), expanded and matured over much of the southwestern highlands in the period between 1950 and 1975 and into the early years of the Ethiopian revolution's socialist development policies (1974–1991). In that latter era, however, agricultural marketing policies decreed by socialist urban technocrats squeezed local coffee farm profits by imposing price controls on coffee and most cereal crops (excluding maize, the cheapest and least marketable of food crops). This state intervention into farmer choice, unprecedented in Ethiopian history, brought a decisive response from coffee farmers that paralleled the Venetian peasants' strategies of three centuries before. Peasants systematically abandoned the state-controlled crops when it seemed more expedient to cultivate maize.

By 1980, state manipulated coffee prices and rising food costs had undermined the balance of the coffee-maize complex. Coffee farmers began to cut coffee bushes and broadcast maize into those plots to avoid state control over coffee marketing. Government policies to resettle drought-affected farmers, recruit forced labor for road construction, and control marketable food crops provided further incentives for farmers to resort in the short term to maize. What had been an agrarian production system seemingly immune to the inducements of maize suddenly rushed to embrace it. In the half decade from 1986 to 1991, maize as a percentage of total national crop production rose from 32 percent to 48 percent.[38]

Maize production also exercised a perverse appeal for peasant farmers outside coffee-growing regions, including the agronomically conservative north. Unlike traditional crops—sorghum, teff, wheat, and pulses—maize offered expediency and higher yields on plots that were increasingly smaller because of the pressure of population growth and land reform. Maize's low labor requirements also proved attractive, as farmers under pressure to supply forced labor for government projects or submit to military conscription could broadcast maize seed after only one plowing, as compared with four plowings for other cereals. Little weeding was necessary for local maize types; they grew well on newly deforested land; and the quick maturation of maize and relatively high yields in proportion to outlay of labor and land (as compared with teff or wheat) meant that there was at least one source of food to sustain families hard pressed by shrinking plot sizes and growing population pressure. The rapid rise in maize production nationally in the years of the revolution (1974–1991) showed the result of farmers' perceptions of both the demographic crisis and the oppressive national government. In 1970–1971 maize had made up 15.3 percent of national cereal production (fourth of all cereals); in 1983–1984, at the height of a famine and after ten years of socialist agrarian policy, maize constituted 28.3 percent of the total cereal crop, first among all cereals. By 1991 more than half of all cereal produced in Ethiopia was maize, an astonishing transformation.[39] The percentage of maize in the nation's overall cereal production has increased steadily since, reaching a dominant position in many small farming systems that have adopted the fertilizer and improved seed packages in the late 1990s.

Seeds of Subversion, Seeds of Industry

Ethiopia and Italy have traveled distinctive paths in their conversion to maize in the late twentieth century. But their recent convergence throws into relief their earlier dramatic divergence. In sixteenth-century Italy, peasant farmers seized the moment early on to embrace maize as a source of food and as a crop that helped break the control of elite classes over agricultural practice. By the mid-nineteenth century, the sumptuary habits of the Veneto's rural folk so depended on maize that vitamin deficiency and the resultant pellagra came to symbolize this dietary monopoly and the lamentable equation of maize consumption with poverty. The triumph of maize production in northeast Italy in the mid-twentieth century paradoxically spelled the end of maize as a peasant crop that had sustained smallholder production and was the crux of a peasant challenge to elite control. In the years between 1947 and 1950, and into the 1960s, maize expanded directly with the importation and development of new hybrid seeds (virtually all yellow dent types) that derived from international markets and post–World War II aid programs led by the agricultural scientists and seed companies preaching the gospel of agricultural modernization. Under this revolution of agroindustrial modernity, productivity more than doubled on Italian maize farms between 1950 and 1960.[40] By 1990, Italy annually produced 6.4 million metric tons on 0.82 million hectares with an average yield of 7.8 tons per hectare. In this new configuration of maize production, however, only 15 percent of the product produced was for human consumption (in the form of flint maize for polenta). Meanwhile in a historical irony, polenta (maize porridge, the old peasant staple) has become an elite food and one of Italy's contributions to international haute cuisine, a far cry from its past as peasant gruel.

By the early 1990s Italy had become the world's thirteenth lead-ing maize producer; in total production it ranked just below South Africa and just above Canada. It has a modern national seed indus-try (though operated by international corporations, such as the Iowa-based Pioneer Hi-Bred Seed Corporation). As northeastern Italy has industrialized, so has its maize production. Only 15 per-cent of Italian maize is the flint variety intended for human con-sumption; the bulk of it is used to sustain the modern dairy and meat agroindustries.

Now in the early years of the twenty-first century, the Veneto's rural fields sport plots of hybrid maize alongside neatly trimmed vineyards that supply low-end wine for local markets. Luxury products of Italian agroindustry, such as Parmigiano Reggiano and Parma ham *(prosciutto),* are both indirect beneficiaries of the Veneto's maize revolution.

16. Maize and grapes, Veneto, 2000.

By contrast, Ethiopia ranks far down on the global list in maize production. In 2002 it ranked twenty-third in the world in the calories in people's diet, but its place on that global list is rising rapidly. As in the overall Africa pattern, 95 percent of all maize goes to human consumers in Ethiopia. Ethiopia's highland producers initially resisted planting maize as a field crop, in choosing rather from the wide range of local grains. Thus Ethiopia's maize revolution has come late and has been one aspect of a decline into precarious subsistence and increasingly centralized control by the state and international agencies, which now seek to break traditional cropping patterns. The grace that maize bestowed on Italy has not fallen to Ethiopia, where maize has been a far more mixed blessing.

In both cases, however, the long-term result has been the globalization of maize in an industrial economy that requires inputs (hybrid seed, fertilizer, and pesticides) and a market infrastructure that link farm production directly to global markets and the erosion of smallholder prerogatives. Seeds from the *mundus novus* thus radically transformed these two Old World political ecologies, but in ways that were quite divergent.

How Africa's Maize
Turned White

5

Just as southern Africa's modern history of mining, industrialization, European settlement, and male labor migration distinguished it from much of the rest of the continent, so did the history of maize take a somewhat different tack there. Colonial Africa's own transformation of the world geography of maize turned African maize white, perhaps the most distinguishing feature nowadays of the continent's maize.

The story of maize in southern Africa offers two contrasting narratives, in which the crop plays a connective role in both theme and substance. One story line concerns the place of maize in the emerging mining and industrial economy of South Africa, and the region as a whole, where the crop in its incarnations both as "mealies" flour and as the main ingredient in beer for urban shabeens (beer halls) fed and entertained mine workers cheaply. The second narrative concerns the small farms, often managed by women, that provided food to sustain workers' families and supplied the bags of mealies that fed men in the cities and mines. Where these smallholder farms did not directly sustain miners or urban labor, they were engaged in a struggle between subsistence farming and cash cropping (usually of maize, tobacco, or cotton) in competition with large commercial agricultural estates that commanded the best land and produced for a national market controlled, often futilely, by national maize boards.

Though the individuality of African maize expressed itself in very different ways on small farms and on large estates, the plant's gradual transmutation after it arrived from the New World displayed many colors and characteristics. What emerged finally was its modern hybrid form, complete with homogeneity of color (white), set needs (nitrogen fertilizer), and an established role to play as the food staple in black diets. Between these two parallel story lines lie many variations, in Malawi, Zambia, Botswana, Mozambique, Lesotho, and other settings. Rhodesia / Zimbabwe is a case in and of itself, since it combined a history of white settlement with active engagement in the modernization of maize along lines as harsh as in South Africa. The story of maize in southern Africa as it appears here is more often anecdotal than comprehensive, but even an attempt to describe the major themes and offer some generative vignettes may help to foster further research into the local African maize stories.

Maize Arrives in the South

Maize made an early appearance on southern African farms and in southern African gardens, not as a leading protagonist but as a supporting player, as it was able to fit into seasonal cycles or specific soil niches between the older staple crops or alongside other New World émigrés such as cassava, potatoes, beans, sweet potatoes, tobacco, or squash. The historical evidence for the arrival of maize in southern Africa, as elsewhere on the continent, is tantalizing but vague. As with West African documents, the historical sources often offer confusing use of the Portuguese terms *milho* or *milho zaburro* for maize and for those two indigenous African crops (sorghum and millet) which dominated southern African food supplies until the early twentieth century.[1] In some cases, however, the references to maize's special characteristics and agronomy are clear. As mentioned earlier, Jan Van Riebeeck did not report seeing maize on the Cape when he arrived in 1652. By 1658, however, he recommended the sowing of maize brought from West Africa's Guinea Coast to the first generation of Dutch farmers. The nineteenth-century historian D. Moodie wrote:

> The Commander at the same time [1658] visited the farms of the free-men, and, as the season for sowing Dutch grain is past, he recommended that each farmer should sow a quantity of *mily,* or Turkey wheat, brought from Gunea [*sic*] by the Hasselt [a merchant ship], as this was the right season and as the slaves were acquainted with the mode of cultivating it.[2]

The reference here to Dutch trade with the Upper Guinea coast strongly suggests that 1658 marked the first arrival of maize on the Cape and that it was by that date already well established in West Africa, and already linked with the slave trade. That slaves (brought from elsewhere in Africa) were familiar with the crop also suggests that maize was already well established along the East African coast.

Two eighteenth-century accounts from European visitors to the eastern Cape recorded the terms *Boona,* and *umbona,* the modern Ama Xhosa word for maize. John Ayliff commented in the early nineteenth century on foods eaten by the Xhosa, describing rather clearly flint maize as a vegetable that complemented sorghum: "Umbona, or Indian corn, which they use when green, by roasting it on the heads, in the embers; and when ripe, it is knocked off the heads, and dressed like the Kaffir corn [sorghum]; and they also eat it parched in hot ashes."[3]

Farther away from the Cape the evidence on the arrival of maize is less clear. W. J. Burchell, who in 1812 visited what was to become the British protectorate of Bechuanaland, made no mention of maize in his list of crops cultivated by the Tswana, though he mentioned its use and cultivation on the eastern Cape at a mission station in 1811; he noted, however, that "maize, or Indian corn was cultivated on the Cape for the poultry." In the early twentieth century the South African agronomist Joseph Burtt-Davy met a

"native" at John Moffat's old mission camp who recalled introduction of maize by missionaries. A centenarian told a certain Father
Norton in 1910 that mealies appeared in Modderpoort district together with the missionaries. Allister Miller of Swaziland reported
to Burtt-Davy that maize was introduced to Swaziland about the
time the Hlamini clan, the Swazi royal clan, crossed from
Tongaland, at the end of the eighteenth century. Reportedly, they
did not call it by the SiSwati name for food, *mabela,* but by the
Zulu words *m'lungu,* meaning "white man," or *m'bila,* the Zulu
name for maize.[4]

Our best evidence for southern Africa comes, like that for the
rest of the continent, from the names and types of maize found already in place at the opening of the nineteenth century. Brazilian
coastal flints, such as the orange Cateto variety or a blue flint (now
often called blue Zulu), perhaps the first maize imports to southern Africa, adapted well to the drier areas favored by South Africa's crop- and livestock-raising Nguni peoples. Early Portuguese
coastal trade contacts or Arab Indian Ocean networks probably
introduced these Brazilian types along the Natal coast or in and
around Delgoa Bay, where inland trade networks brought them to
landscapes whose farming systems had previously mixed livestock
with sorghum and millet. If the yields of these short-season hardy
flint types did not challenge drought-resistant sorghum, fresh flint
maize was edible earlier in the hungry season than sorghum and
could stay longer in the field without suffering damage from birds.
Later, soft floury maize types arrived, as they did in West Africa,
possibly also from Brazil, probably by way of Angolan and Mo-
zambican trade networks. Dent maize, the type that served as the
basis for the hybrid-maize revolution in the Americas, did not arrive in southern Africa until the very end of the nineteenth century, but then it had a decisive effect on the course of events in agronomic, nutritional, and political terms. The arrival and quick
adoption of dent maize reflected South Africa's emergence as a site
of mining, industrialization, and urban growth.

In areas beyond the Limpopo River, the penetration of early flint maize moved in tandem with the expansion from the east of the Indian Ocean economy. There is one claim that maize was already a staple in northeast Zambia's Bemba-speaking Luapula district by the end of the eighteenth century, though maize economist Marvin Miracle argues, probably correctly, that maize predominated nowhere in the region until the beginning of the twentieth century.[5] Nonetheless, the linguistic evidence from the region strongly corroborates the maize plant's local association with the Indian Ocean trade—that is to say, establishes it well before its spread within South Africa itself. That Nguni languages, such as Zulu, share variations of the word *umbila* (or *mapira*) with a number of East African Bantu languages implies movement of the words and of the plant itself from the north and east.[6]

Portuguese colonies in southern Africa seem to have had maize earlier than South Africa itself did. As early as 1590, the Englishman Andrew Battel, who had lived in Angola for eighteen years, reported that the local population was cultivating "the great Guiney wheat, which they call *mas-impota* [*masa muputo* is a modern name for maize there]."[7] By 1821 maize was already well entrenched as a field crop in the area of Mozambique around the colonial capital Maputo (Lourenço Marques). By the middle of the nineteenth century in landlocked southern Malawi, maize—or *chimanga* (from the sea) in Chichewa and other languages of the area—was already part of a hand hoe–based, woodland fallow cropping system in low population areas where *citamene* (shifting agriculture) was the norm and when maize functioned as a short-season garden crop.

Maize then spread into the Shire highlands along with new varieties of cotton and tobacco that reflected the growing regional exchange economy.[8] In drier areas of the region, where finger millet and sorghum were still the dominant cereal staples, farmers planted the more moisture-sensitive maize seed (mostly flint varieties) in moist alluvial patches in riverbeds (*dambo / dimbe* land), which they then covered with sand to prevent both waterlogging and evaporation. Malawian farmers, in fact, preferred these local flints to more "modern" dent varieties until late in the twentieth century. On the farm, Africa's early flinty maizes probably also appealed to women, who preferred flint maize's high grain-to-flour extraction rate because the flint germ separates from the bran more easily in the mortar than it does for higher-yielding dent types. The longer husk of flint maize also protected the ears from weevils in the field and during storage.[9]

17.
South African flint
maize types, ca.
1910.

The first generation of colonial ecologists who wrote in the early twentieth century managed to understand the farming systems that emerged out of the nineteenth century and offer a good view of what preceded colonial rule. William Allen, deputy director of agriculture in Northern Rhodesia, for example, was a sensitive observer of local agricultural practice on the eve of central Africa's maize revolution. He described the classic central African "ash garden" cultivation found in much of Malawi and parts of Zambia that blends New World and African plants in a system that balances plant physiology, soil chemistry, and dietary diversity. Maize was from early on only a small part of this complex mix: development of the "ash garden" from lopping and brush collection involves the subsequent rotation of finger millet for the first year and groundnuts in the second year, and in the third year the garden is "thrown into mounds on which beans are grown and intermixed with other crops such as maize, cowpeas, gourds, pumpkins, and sweet potatoes . . . [Farmers] have evolved or adopted an ingenious system of alternating maize and beans on mounds with finger millet planted on a seedbed prepared by breaking down and spreading the mounds."[10] Maize in these conditions was a vegetable crop that fit neatly into a niche in household food supply, if not in the regional exchange economy, a role it continued to play during the first third of the twentieth century.

When British anthropologist Audrey Richards carried out her pathbreaking study of nutrition in the Bemba area of northern Zambia in the 1930s, she caught a society in transition from an agrarian to a modern economy, the latter characterized by labor migration and new diets. She offers a telling account of the change in diet and perceptions of food:

Bemba, after leaving their country to work in urban areas in the south, say they find it difficult to adjust themselves to the maize flour "mealie meal" they are given there. One old man probably too fixed in his gastric habits to become adapted to town life said, "Yes, first I ate through one bag of [maize] flour and then a second. Then at last I said, 'Well, there it is! There is no food to be found among the Europeans.'"[11]

The old man's complaint about maize as unsatisfying "European" food is reflected in his obvious disdain for the new food of the mines and cities of colonial Northern Rhodesia (Zambia). But the derision he expresses for the newfangled food probably applied equally to a set of rapid changes in the economic and agricultural landscape of his world, in twentieth-century Africa in general but in southern Africa in particular.

Perceptions of maize as "European" food were anachronistic even in 1939, reflecting a landscape of memory rather than the creep of agrarian modernism. Zambia's first president, Kenneth Kaunda, who was born in colonial Nyasaland in 1924 but grew up in Northern Rhodesia's Bemba-speaking Northern Province, recalls that maize was ubiquitous, but only as a garden crop.[12] Over the course of his lifetime and that of his generation in southern Africa, maize came to dominate both the national diet and the region's international politics.

In fact, by the 1930s diets and farm plots all over southern Africa were well in advance of movements and meaning of maize in the rest of the continent. Southern African farms and farmers were rapidly changing the seasonal tastes, textures, and colors of their diets to reflect the inexorable spread of a food crop with origins in the New World but with an increasingly African identity.

If Africa's original maize imports were flint maizes that fit ideally into farm niches as a vegetable, the arrival and rapid spread of American dent types was in many ways indicative of a shift from a mercantile era to the industrial age of grain as commodity—an economic leviathan. Maize in that earlier era was an element of biodiversity and farming as an eclectic art, which would give way to homogenization and an expansion of global scale.

New Dents for Old Flints

South of the Limpopo in South Africa another, quite different maize-led transformation took place in the late nineteenth and early twentieth centuries, as a part of the mining revolution and industrialization that spiraled out from the demands, in 1867 at Kimberly (in the diamond mines) and in 1886 at Witwatersrand (in the gold mines) for food supplies for mine workers. The evolution of a distinctive agricultural landscape, in the regions of the Transvaal, the eastern Orange Free State, and colonial Basutoland (Lesotho after 1968) known as the Maize Triangle, was a product of the historical conjuncture of maize, labor migration, and nascent industrial capitalism on the South African highveld. This setting of highlands, palatable grasses, friable soils, and well-defined rainfall frontiers became a stage on which maize played a seminal role in struggles over labor, landscape, and livelihood. Maize stood at the heart of an industrial transformation of the national food supply.

Events of the mid-nineteenth century transformed the Vaal and Caledon River valleys on the highveld into space for new demographic expansion and agricultural innovation. In the wake of the Zulu Mfecane expansion of the 1820s and its dispersal of peoples on the highveld, as well as the Cape Dutch Great Trek migration from the west after 1838, new populations and old pushed forward competing ideas about landscape and land use. For the Basotho people caught in this demographic crucible, the highveld landscape was a known quantity in a farming system relying on the hand hoe and on sorghum as the primary grain crop, but it was terra incognita when it came to adjusting to the unfamiliar crops and tools that arrived as baggage of southern Africa's emerging political economy. The tools included the heavy moldboard plow, use of oxen as draft animals, and the introduction of wagons and sledges for transporting bulk items (such as bags of maize) to distant markets.

By the mid-nineteenth century, maize was already entrenched as a part of the highveld mixed cropping system, albeit a rather junior partner.[13] In this era, sorghum was the preferred food and the dominant crop, though flint maize appears to have been one of several complementary crops sown in fields that women prepared with hand hoes and weeded cooperatively. The friable, sodic soils required little clearing or plot preparation, so Basotho farmers could quickly make their escape to defensive positions in the foothills at any sign of an enemy. Eugène Casalis, a French Protestant missionary, described in his diary his first encounter with Basotho at the western edge of their territory in June 1833:

> We reached the foot of a beautiful mountain [Thaba Nchu] five to six hundred metres in height and several kilometers in circumference. Directly under the mountain, we could see large fields of near ripe maize and sorghum (large millet). The inhabitants had built their huts on the steepest summits, as a precaution against enemy attacks. Those who were busy in the plantations fled at our approach.[14]

By the 1830s, the Basotho had moved back to the lands of the lower pediment, to reclaim cultivable plots, some of which offered abundant winter lowland grazing on *Themeda trianda* grasses. Maize in this setting already appears to have been well established within the local diet, a food that refugees could harvest early, roast quickly, or prepare at a milky stage well before the season's planting of sorghum was mature. In 1840, Thomas Arbousset, accompanied by the Basotho chief Moshoeshoe and his son, visited the chief of an isolated village tucked into the Maluti Mountains:

> The good man offered us some sweet reeds; then his wives arrived to roast some corn cobs; from time to time they also brought us a

kind of moist bread or *bohobe*, made from half ripe Indian corn, from Turkish corn, and pumpkin. All this ground and boiled together without salt is not, it is true, very appetizing; but what does one not eat when one is hungry?[15]

The half-ripe maize served to the guest and described in other accounts of the period was undoubtedly the green milky stage of the older flint type maize that people had carried with them as, fleeing from the violence of the Mfecane, they escaped from the highveld and into the mountains. As in the case of the moist bread, again, maize was served as a vegetable complementing sorghum (the preferred staple) and squash, rather than being the grain staple it later would become. Also, unlike sorghum, whose cultivation required constant vigilance against bird depredations, maize could be left in the field in otherwise open, dangerous areas and harvested in stages as safety permitted.

The discovery in 1867 of diamonds at Kimberly on the highveld between the Free State and the eastern Cape rewrote the regional economy. Two of the first impacts of the influx of workers and capital to the northeastern Cape were a burgeoning market for food and the birth of an economic infrastructure in the region, including a cash economy, road networks, and new technologies of transport and production, especially wagons and moldboard plows.[16] The response of Basotho farmers to these market opportunities was astonishing: in 1873 alone, Basutoland exported more than 8,000 metric tons of grain, perhaps a third of it maize. Export figures by 1893 had more than doubled, to include 11,600 tons of wheat and 6,000 tons of maize.[17] By 1900 the region's economy reflected the Basotho response to urbanization and regional links to international markets: winter wheat, maize, Merino sheep, and Angora goats superseded Basutoland's agricultural staples of sorghum, milk, and cattle. More subtly, these changes redefined maize, from garden vegetable to commodified grain; new urban consumers ate their maize not as a snack, but as mealie flour, which doubtless came initially from maize of various types and colors that Basotho farmers had collected from local trade and seed stocks.

In addition to market access, trade brought new seeds suited to large-scale production. The arrival of the railroad at Kimberly in 1885 opened the regional markets to cheap grain imports and un-accustomed ideas from Australia and the United States. Sometime before 1898 these contacts provided a critical introduction in the form of a new type of maize: American white dent types arrived in the late nineteenth and early twentieth centuries, bearing such homespun names as Boone County, Iowa Silver Mine, Hickory King, and Horsetooth. The most important of these early imports was probably the medium-late white dent maize known as Hickory King, which was to become one of the most important progenitors of southern and eastern African commercial maize types over the course of the twentieth century. Hickory King produced an eight-

18. Maize drying on a South African black farm, late nineteenth century.

row (and later, ten-row), large-grained dent variety that tolerated poor soils and outyielded the older flint maizes. Dent maize, in contrast to its hard-starched flint cousin, not only had higher yields than older maize varieties and sorghum but also had a soft starch suited to the mechanized mills in mining towns and new highveld urban areas.[18] Dent maize now offered the possibility of maize production on an industrial scale for export and as a cash crop for both highveld black sharecroppers and commercial farms (after the 1910s).

Production on a grand scale made maize in South Africa potentially an exportable commodity, though it was valued in international markets more as an industrial raw material (starch) than as a food grain.[19] Dent maize also provided a direct link to an international revolution in dryland agriculture, centered in the American Midwest, that had begun to leapfrog to Canada, Argentina, Brazil, Australia, and, after 1917, the Soviet Union. This global network expanded even more rapidly after the 1920s, when the genetics of hybridization gave birth to a new type of agricultural science.

The arrival of the ox plow on the veld in South Africa and its peregrinations with missionaries, merchants, pioneer columns, and entrepreneurs onto lands further north transformed agrarian ecologies, as maize shifted from a short-season garden crop to one amenable to large-scale cultivation. Maize was better suited than wheat for this purpose because it grew in a wider range of environments, was simpler to process, required much less labor, and resisted bird damage.[20] Because maize matured to usable food earlier than sorghum or millet it could be planted and harvested at points that allowed households to manage and maximize their labor calendars and family food supply.

The same regional economic boom that fueled the rise of large commercial farms in what came to be called South Africa's Maize Triangle also increasingly drew male labor from neighboring peoples to the mines and away from full-time agriculture. In 1873, for example, fifteen thousand Basotho men worked in South Africa's diamond mines; in 1886 (the year gold was discovered on the Witwatersrand) the number doubled. Male wage labor had thus rapidly supplanted agriculture as a source of cash for rural black folk. The Native Lands Act (1913), a pillar of what was in 1948 to become apartheid policy, resulted in mass dispossession of black

19. South African maize, ca. 1915.

sharecroppers and underwrote white farms' investment in commercial maize production. By the 1920s only a fifth of Basutoland's adult males were still employed in agriculture.[21]

The entrepreneurial economy of black farms in the Orange Free State / Basutoland region that had shown such resilience in the 1870s suffered two simultaneous shocks in the first third of the twentieth century. First, the loss of men to the mines left local agriculture in the hands of women, who now were compelled to manage household farm production without male labor. Second, the feminization of agriculture shifted strategies away from labor-intensive crops (for example, wheat was relatively expensive to produce and was available more cheaply from overseas markets) to low-labor maize. Winter wheat's requirement of timely labor at harvest raised production costs and increased harvest losses. By contrast, maize offered attractive but contrasting properties. It was an early-yielding, low-labor crop (by comparison with sorghum and wheat) that provided food for the family; it was also attractive to labor-poor farms because it could remain in the field until seasonal labor costs declined, and farm families could then harvest their own crop.[22] As labor costs forced wheat farms into mechanization to survive, maize farms, by contrast, could make effective use of off-peak labor.[23] For those large-scale white farms with access to capital, improved dent maize—especially Hickory King—

20. An old-style eight-row Hickory King corn ear, 2003.

21. Mechanization of South African maize cultivation, ca. 1910.

22. White farm, black labor: mechanized maize processing in
South Africa, ca. 1910.

was also a high-yielding cash crop amenable to commercial fertilizer, improved seeds, and industrial-scale production. Maize was thus highly compatible with both the national labor equation that called for mine workers *and* the need for a cheap food supply.

It seems, however, that black South African farmers embraced the planting of maize, mainly floury and flint, on their farms long before white settlers did and well in advance of dent maize's commercial debut. Even at the end of the nineteenth century, white South African farmers still argued that maize was a "Kaffir" crop with little commercial value. In the Orange Free State, South Africa's largest maize-growing area, only 2.25 percent of the area in the province was planted to the crop in 1914, mostly by black sharecroppers.[24] Only with the arrival of the American white dents and their local descendants (such as Potchefstroom Pearl and Salisbury White) did white farmers take to maize as a major cereal crop, and then only after black African farmers had begun to exploit its usefulness and profitability in their quest to supply food to growing mining centers. Like the Colt .45 revolver that won the American West, Hickory King and other American white dent maize provided the agrarian economic base for the rapid expansion of settlers' rule in southern Africa.

Maize had as great an effect, or greater, on southern Africa's black population. It was no accident that by 1930 maize had surpassed wheat as the region's major cash crop and had replaced sorghum as the major food crop. The seasonal exodus of male labor from Basutoland meant that the former food-exporting colony had also become a net grain importer.[25] For Basotho on the highveld, maize fed both sides of the "divided family": males in the mining hostels ate commercial mealie papa (porridge), and women and young children on impoverished farms on the black homelands consumed their local crop. It was those bags of cheap mealies which after 1948 continued to underwrite the economic foundation of apartheid.[26]

In the late nineteenth century and the first third of the twenti-

eth century few differences were evident in the agricultural economies of the political units within the Maize Triangle. The Orange
Free State, the southern Transvaal, and Basutoland each served
the growing industrial economy with labor and food. Orange Free
State farmers, whose land was more arid, developed livestock
ranches, whereas Basotho areas that received at least five hundred
millimeters of annual rainfall concentrated on grain production. By
the 1930s and 1940s, however, the paths of their rural economies
crossed and took very different trajectories. Black farms had ceased
to produce food crops for the market and now relied on maize for
subsistence.

White farms, by contrast, had grown in scale and attained access
through price controls and government programs to resources of
credit, extension, and markets. Highveld mining towns and marketing centers like Bloemfontein, Maseru, Harrismith, Bethlehem,
and Potchefstroom organized both the market for maize and the
distribution of new crop varieties and equipment. The introduction
of new high-yield, open-pollinated dent varieties like Hickory King
and Potchefstroom Pearl, as well as industrial milling (which preferred dent maize to flint) in the new towns and cities encouraged
such monocropping and, eventually, the mechanization of maize
farming.[27]

What's in a Color?

Within the twentieth century, as maize came to dominate southern
African politics and diets, a phenomenal and very visible change
took place in maize itself, reflecting the influences of science, African choices, world markets, and modernity. Despite the wide variation in localized tastes, traditions, and aesthetic choices among
African farmers and consumers, historical serendipity, and the ec-
centricities of maize genetics, African maize is, at the beginning
of the twenty-first century, overwhelmingly white in color. White
maize is what Africa's farmers produce and sell; and white maize is

overwhelmingly what African consumers eat. Africa's almost un-paralleled, and very recent, color choice makes the continent stand out from other areas of the world. World production of white maize is, however, less than seventy million tons, its production dwarfed by the global output of more than five hundred million tons of yellow maize. Yet white maize makes up more than 90 per-cent of Africa's total maize crop, and Africa produces about 33 per-cent of the world's white maize. In eastern Africa virtually 100 per-cent of marketed maize produced is white in color.

In West Africa, the presence of other varieties and colors, so striking in earlier times, has now largely receded: 90 percent of the maize West Africa produces is white. Within Africa, only in South Africa does yellow maize appear in production statistics, and there only because of that country's commercial livestock and poultry in-dustry. In South Africa, maize for human consumption is almost exclusively white.[28] Central America is the only other world area that shares Africa's passion for white maize; it is also the only other world area that consumes more maize as human food (95 percent) than for livestock and industrial purposes. Whereas Central Ameri-can consumers have developed their preference for white maize over the long haul, since its domestication some six thousand to ten thousand years ago, African consumers accepted white as the color of their staple crop only beginning in the 1920s and 1930s. Therein lies a story.

The African preference for white maize is firmly held, but based primarily on a relatively recent aesthetic bent, for many have ar-gued that blind tastings show no real difference in flavor between kernels of different colors. Chemically and genetically, white maize is virtually indistinguishable from yellow maize. The difference in appearance of white maize stems from the absence of the caro-tene oil pigments responsible for the color of yellow maize. Nutri-tionally, yellow maize contains traces in its seed coat of vitamin A, while white maize does not, though the amounts of the vitamin in the yellow type fall well below daily human requirements. In livestock feeding, yellow maize has a preferred value because it gives egg yolks, poultry meat, and animal fat a yellowish tinge preferred by consumers in many cultures. Yet Italian consumers, for example, also prefer white maize for polenta; in the Veneto and most of the rest of northern Italy consumers favor white polenta with fish and yellow with meat. Farther south, the opposite is true—though one is hard pressed to find any polenta at all in shops or restaurants south of Rome, where pasta is king.[29]

Despite the recent evidence that African consumers are now rather firmly entrenched in their preference for white maize, earlier Africans' aesthetic sensibilities regarding food and symbol caused

23. Africa's maize rainbow: local varieties of central African maize, 2003.

consumers to select colored maize, ranging from crimson to blue to colorful mosaics of red, blue, yellow, and orange, either on different ears or all mixed on a single ear. Historically, many local consumers in Africa expressed strong preferences for red or blue or orange or yellow kernels, or for variegated mixes of all of the above.[30] In the early 1960s the presence of colored maize on African markets began to recede, as national trends favoring white maize overtook older local traditions.

In some parts of Africa—notably those like Angola which are less thoroughly invested in maize as a commercial crop—yellow retains its appeal for consumers. And in most areas smallholders retain seed for the early-yielding and brightly colored flint and semiflint garden maize.

The historical color bursts in older types of African maize doubtless reveals its genetic diversity and the several points of New World origin in trade, conscious experimentation, and serendipity. Loss of color doubtless also reflects loss of the biodiversity that traveled from the New World to the Old in the years after 1500. The diversity in its germplasm is an asset under threat from popular and economically attractive modern hybrid varieties that have been manipulated for yield and uniformity of color.[31]

How did Africa change from being a repository of maize genetic diversity in the first four centuries of its encounter with the crop to promoting an almost complete homogeneity of color, or lack of any pigment, by the end of the twentieth century? The answer would appear to be the maize's transformation (and industrialization) from farm-consumed vegetable to marketed, urban food in the form of flour, a process that began early in the twentieth century and largely in southern Africa. In the late nineteenth and the twentieth centuries the colonial economies of southern Africa experienced a simultaneous demand for food supplies for urban areas and mining centers, a demand met largely by maize flour. At the same time, colonial economists in South Africa and the Rhodesias saw maize as their most exportable agricultural commodity, especially for the British industrial starch and distilling market. Each of these sources of commercial demand put a premium on uniformity in quality and appearance. Indeed, in 1911 the secretary of the London Corn Exchange informed South African and Southern Rhodesian commercial maize farmers that southern African grain could compete on the European market only if it improved its grading and uniformity.[32]

Because the British markets paid a premium price for white maize over North American yellow, and mixed fields of white and yellow tended to cross-pollinate and to yield variegated ears, colonial agricultural law undertook systematic management of maize color and types. In 1921, for example, the Rhodesian Maize Seed Association passed a resolution stating that the growing of yellow maize in the territory constituted a "vital danger to the maize growing industry." In Southern Rhodesia the Maize Act of 1925 allowed commercial farmers to restrict the growing of maize in their area to particular types and colors (namely, to white dent as opposed to flint and any other color).[33] For the new maize mills supplying mines and urban African markets, the concern with uniformity was similar, and the commercial millers found the soft dent types vastly preferable in both uniformity and texture. Even if African women who hand-milled maize preferred the way flinty maizes yielded more flour per bushel and people still liked its flavor and texture, export-minded commercial farmers opted for the yield and consistency of white dent. These commercial farmers had the political clout to win the day, and Africa's maize made an inexorable shift to white.

The subtle struggle over maize's dominant color took place most explicitly in southern Africa. Writing in the early twentieth century, South African maize scholar and entrepreneur Joseph Burtt-Davy offered a view of the question of maize color that is both insightful and peculiar to South African racial thinking. He details the varieties of maize imported from America as well as the contest between yellow and white, white and black. In 1914 he wrote:

For the mills supplying the Rand mines, the large flat, white grain produced by the Hickory King (8-row), 10-row Hickory, Hickory Horsetooth, Mercer, Ladysmith and similar large-grained dent breeds, is in the greatest demand when a choice is offered. This is partly due to "trade fancy," but some millers state that there is less bran produced in milling these sorts than is the case with the small grain. Where there is no choice of flat whites, any flat white dent is acceptable to the miller in preference to yellows or even to round whites. At one time, the writer is told, the natives employed on the mines would eat yellow mielie meal in preference to white, but now it is the exception for a Rand native to eat anything but white meal. Various excuses are given—such as the undoubted difference in flavour between the white and yellow meal; the supposed injurious effect of yellow meal on the digestive system, etc. But in view of the large amount of yellow "corn meal" consumed in the United States, one is scarcely prepared to accept these as valid reasons; it seems more probable that the real cause is the tendency of the native to imitate the white man, and that as the white man in South Africa eats only white mielie meal, the native thinks he ought to do so too. The reason may also be partly commercial; millers prefer to mill only one colour of maize, and may have been instrumental in gradually inducing the mine natives to use white meal, not only for that reason, but also because white maize is usually cheaper than yellow in the Johannesburg market.[34]

At the same time, early maize breeders sought to improve southern Africa's maize yield and to achieve uniformity of the crop in the field. They consciously sought new varieties locally and from the New World suited to the market and to industrial needs. By the 1920s, by law and by market preference, white dent maizes had become the standard crop of any farmer who sought to engage the market, particularly for export. Flat whites (or dents) later became the foundation for both the wildly successful SR-52 in Southern Rhodesia and the Kenya Flat White complex that dominated African maize breeding in the later twentieth century. Colorful flints, older dents, and other manifestations of the original genetic diversity of Africa's first maizes continued to exist, but only on small farms, where they served as a vegetable crop or to satisfy farmers' nostalgia.[35]

But what accounts for the preferences of farmers and consumers outside southern Africa, since the data show that the predominance of white maize affects most of West and eastern Africa as well? The answer probably lies in the powerful influence southern Africa wields over research on improved maize, the desire of urban markets for uniformity in food supplies, and a pervasive drive toward modernity on farms and in fashion. Farmers inevitably plant and harvest small amounts of their colorful old flint maize and even some semidents in red, orange, or yellow. But they market little of this, and local grain merchants give strong price preference to uniformity.

Having left behind a venerable rainbow of colors and types, African farmers and consumers have switched the debate to white versus yellow (the latter dominates world maize markets). In the southern African droughts of the 1980s, governments tried, and largely failed, to convince consumers to accept imported yellow maize. In fact, southern African consumers readily pay more for the privilege of eating white maize rather than yellow. In Mozambique in the mid-1990s, for example, the price premium for white maize grain was up to 25 percent, for white maize flour over yellow it was 15 to 30 percent. Among Zimbabwe's urban consumers in mid-1993, poor households paid a 33 percent premium for white flour, while high-income urbanites willingly paid 85 percent more for white flour over yellow![36] My own informal survey of Harare black-market mealies outside a Harare supermarket in 2003 showed a 15 percent markup for white maize flour over yellow. Government policy in much of southern Africa, in fact, stockpiles white maize to avoid the signaling of a food crisis implied by the appearance of yellow maize on local markets.

Beyond the premium price, Africans also pay an additional hidden cost because the productivity of white maize lags behind that of yellow, primarily because there has been less research in industrialized countries into improving white varieties. In the United States, for example, yields for white maize are 10 to 15 percent below those for yellow. In 2003 research had only just begun on a genetically modified (GM) white maize in Kenya and in the United States. Moreover, the cost of processing white maize (largely intended for human consumption) is higher than for yellow maize because the latter is usually processed less carefully as livestock feed. This difference in processing may, in fact, account for consumers' insistence that the white type is of better food quality.[37] The influence of southern Africa, as a market and as the site of new research, on Africa's maize crop will continue in the near future.

The Measure of History

At the end of the twentieth century, political and economic forces inscribed two extremely divergent landscapes of Africa's southern regions that shared maize as their dominant crop. White farmers there took advantage of a postwar boom and their political domination to invest heavily in large land units, cheap black labor, and the technology of dryland production of winter wheat and hybrid maize. Since the 1950s, government subsidies from South Africa's mine-based economy have allowed white farms to invest in tractors, harvesters, antierosion measures, and postharvest processing already tested on dryland farms in the United States and Argentina.[38]

South Africa's expanding Maize Triangle and the Zimbabwe highveld have, in fact, now joined the international maize industrial complex. From its modest origins based on the coexistence of flint maize with traditional cereal crops, maize has developed into southern Africa's most important cereal crop. In 1999 maize occupied three-fifths of all land planted in cereals. It represented two-thirds of total cereal production in South Africa and more in other nations of the region.[39] Mealies, consumed primarily by black South Africans, provided more than a third of the nation's carbohydrates, a sixth of the fat, and three-tenths of the protein in local diets. Since 1994 South Africa has received 441 new maize breeding lines from the International Center for Maize and Wheat Improvement and has now developed its own private seed companies to market hybrid lines to large-scale farms. Unlike markets elsewhere in Africa, however, South Africa's developed meat markets consume a quarter of its maize (all yellow) as livestock feed.

The black farms of Basutoland and South Africa, by contrast, emerged from the post–World War II years mired in the rural economy of labor reserves that provided male workers to mines and commercial farms: farms deficient in labor and capital produced maize primarily for subsistence. The networks of erosion gullies that zigzag across the countryside in Lesotho and on the eastern Cape, collecting rusting auto carcasses and swallowing farmer's maize fields, are only the most visible manifestation of a fall from grace. Archival records from the 1920s and 1930s show that in colonial Basutoland the incidence of pellagra and kwashiorkor increased in direct proportion to the rising percentage of maize in the local diet and the decline of sorghum as a staple during the postwar years.[40] The Lesotho economy, which had been a major grain exporter in the late nineteenth century, was by the mid-twentieth century a net maize importer. Maize, which had been an engine of its growth, eventually became a measure of its decline.

African Maize,
American Rust

6

In September 1949 F. C. Deighton, a colonial
plant pathologist at Sierra Leone's Njala research station, reported with alarm the presence
of a brownish red fungus on the leaves of that colony's maize crop.
He thought it to be a variant of the fungus *Puccinia sorghi* and
called the new form American rust. The damage in that crop year
in Sierra Leone was "slight." But his diagnosis had been mistaken.
His American rust was a disease new to West Africa and the Old
World and was about to bring about devastating crop losses in Africa and around the globe in the coming years.

Within three years of its first appearance in Sierra Leone this
highly virulent maize fungal disease had spread throughout West
Africa's humid zone to the Gold Coast, Ivory Coast, Nigeria, and
then across the entire African Maize Belt from west to east. As the
disease moved across West Africa in 1950 and 1951, its effects
grew from "slight" to severe, and it destroyed as much as 50 percent of vital maize harvests. In 1950 the price of maize in some areas of West Africa rose by 500 percent. By 1956 the disease had
spread beyond Africa and sped across the globe's southern hemisphere, infecting maize in Madagascar, Australasia, and as far east
as Christmas Island in the Pacific.

Beginning with the 1949 report from Sierra Leone, colonial officials mobilized a network of applied scientists and research insti-

tutions linked within the British colonial administration to respond to this immediate threat to the food supply of their colonial possessions. But shortly thereafter, the limits of the colonial science infrastructure gave way to a new postwar reality. By the height of the American rust crisis in 1952, crop scientists, geneticists, and plant breeders from around the globe had begun a frenetic effort to identify the fungus, to find resistant maize germplasm, and to set up breeding programs to solve the threat to colonial food supply and economic stability. Joining practitioners of colonial science were powerful new American institutions, and a new array of multilateral agencies, progeny of the postwar world. By 1953 this self-confident global network had produced new strains of rust-resistant maize but found to its surprise that the threat American rust posed to the world's food supply had vanished as quickly and mysteriously as it had come.

What happened? And what can we learn from this seemingly insignificant and now obscure episode of crop disease? The incident illuminates the process whereby maize became a part of the institutional life of colonial Africa as well as the subject of an emerging global development science. The short-lived crisis of American rust helps us understand the nature of science, political ecology, and the globalization of power at the end of Europe's formal age of empires and the beginning of the multilateral postcolonial world. Maize as a crop and as an object of scientific empirical study was part and parcel of the new science of development that emerged full-blown after the 1960s and continues into the twenty-first century.

American rust itself was an intense but short-lived threat to colonial food supply, but the response came from a mature imperial world on its way to becoming a world dominated by the modern development industry and invasive multilateral organizations. In the 1950s the organizations and economic philosophies that would dominate the political ecology of development were only embryonic in form, but their global reach was increasingly evident.

The process of adoption of and adjustments in food ways and farm practice in West Africa was uneven across ecosystems and cultural settings, but maize had clearly taken on an African flavor and local accent. By the middle of the twentieth century West African farmers, processors, and consumers had established distinctive tastes and aesthetic preferences. By 1950, the temporal setting of our story, maize had achieved a prominent status in the African food supply in what had become key British colonies in West Africa (the Gold Coast, Nigeria), East Africa (Kenya), and in southern Africa (Malawi, Northern Rhodesia, and Southern Rhodesia). In South Africa and the Rhodesias the emerging mining economy increasingly depended on cheap black labor fed on *papa, sadza,* and *ishimi* (stiff maize porridge) washed down with maize beer. French West Africa (especially Dahomey, Ivory Coast, and Togo), the Belgian Congo, and the Portuguese colonies of Angola and Mozambique were equally dependent on stable supplies of food at stable prices.

In East Africa the eventual domination of maize as a food crop was in its early stages. In Kenya, where the Mau Mau crisis simmered throughout the early 1950s, smallholder black African farms and large-scale white farms alike depended on maize as a food and a commercial crop. During the 1950s Kenya's farmers relied primarily on the South African maize types Hickory King and Natal White Horsetooth (together called the Kenya Flat White complex) introduced first by white settlers early in the century, and on soft-starch Cuzco types imported from Peru by missionaries early in the century for high-altitude zones (more than 2,400 meters above sea level). Though Kenya's first hybrids did not appear until 1964, maize had by then replaced sorghum and millet as the staple cereal. Likewise, in the Tanganyika protectorate maize dominated the west and highland zones around Arusha, Moshi, Pare, and Usambara. In those places maize had not only taken on the name that had previously referred to the indigenous sorghum *(pemba),* but the newcomer had quickly usurped its place in farmers' fields. In Uganda, colonial officials discouraged maize cultivation in northern cotton areas (because maize reputedly harbored the American boll worm) but *Zea mays* had emerged as a major crop in areas surrounding Lake Victoria, like Busoga.[1] Maize in East Africa thus constituted both a critical food crop for small-holders and a cash crop for large-scale estates on the "white" highlands. During the decade of the 1950s maize production in East Africa increased by 66 percent, from 1,106 to 1,837 tons. In that same period in Southern Rhodesia (Zimbabwe) maize production increased fourfold. Moreover, the trend of maize's expansion into African diets in the decade of the 1950s would continue through-

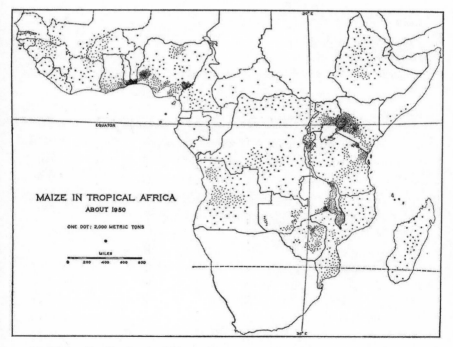

MAIZE IN TROPICAL AFRICA
ABOUT 1950

ONE DOT: 2,000 METRIC TONS

24. Maize distribution in Africa, 1950.

out Africa, and by the end of the century maize would become in some cases a monoculture eclipsing all other food sources (Appendix Table 6.1).[2]

In the postwar decades the European colonial empire was in the process of building urban economies, a stable labor force, and political institutions, all of which relied on stable food prices at the farm gate, at the mine head, and in urban markets. Colonial officials and agricultural scientists were increasingly aware of the local economies' relative dependency on the crop, its levels of production, and long-term trends.

Despite maize's growing importance in much of Africa, however, colonial agricultural research had neglected the crop, instead concentrating research investments on cash crops such as coffee, cotton, palm oil, groundnuts, and cocoa that linked African economies more directly with emerging world commodity markets. It is not surprising, therefore, that at the time of the outbreak of the rust disease no maize specialists were on the staff of the colonial agriculture service in British West Africa.[3]

Rust and Global Scale

The seemingly random arrival and impact of one of the New World maize rusts in Africa in precisely September 1949 is an environmental and historical puzzle. The line of infection that began somewhat innocuously in Sierra Leone moved quickly within the region along the coast to Ivory Coast, the Gold Coast, Dahomey, and Nigeria, where prevailing westerly winds may have carried the aecidospores (transfer spores, the infectious agents) along continuous maize-growing zones from Sierra Leone east to Dahomey and Nigeria. But the rust also spread within two years of its arrival to southern Africa and East Africa, where there had previously been no evidence of it. What accounts for the geography of its spread? The best guess is that the spores, like the Spanish flu virus of 1918–1919, had traveled along emerging lines of colonial communication that presaged the globalization of the late twentieth century.[4]

Rust is the common name applied to any of a group of parasitic fungi that form orange-red spores on the stems and leaves of the seed plants they infect.[5] The genus *Puccinia* has some of the most complex life cycles of any fungi, including several different spore stages; these stages include formation of uredeospores ("summer" spores) as a result of the initial infective aecidospore stage. Reddish brown pustules (thus the term "rust") burst through the host maize leaf epidermis, dispersing in the wind to other maize plants. The fungus does not kill the plant but causes leaf lesions that decrease photosynthesis and other metabolic activities and thus reduce grain yield. The fungus then "overwinters" as dormant teliospores that proceed to go through a sexual phase, to become aecidopores, carried by the wind to infect new maize plants.[6] The rust thus spreads rapidly from plot to plot and from region to region, appearing in each place to be episodic. Tropical rusts infect maize only in humid conditions where the temperature exceeds 80° F (27–28° C) and thus take on a seasonal character consonant with the bimodal seasons of wind and rainfall that predominate in Africa and the tropics in general.

What its first observers called American Rust, thus traveled from New World to Old and from British colonies to French, Belgian, and Portuguese holdings, global paths created by the late capitalist empires. And the rust traveled more quickly than the attempts by colonial agricultural science to forestall its effects. In 1949 maize was only a minor crop in Sierra Leone (rice predominated there), but its moist tropical climate was ideal for rust infection. By the 1950 crop season, however, the mysterious rust had leapt across colonial boundaries and begun appearing all along the West African coast, this time in the Gold Coast, southwest Nigeria, and the French colonies of Dahomey and Ivory Coast, all economies where maize was a central item in the food supply. In Nigeria and the Gold Coast, British colonial officials estimated crop losses to be 40–50 percent and in Dahomey, then perhaps West Africa's most maize-dependent colony, maize prices increased to 124 shillings per hundredweight, a 500 percent rise in a single year.[7]

In spring 1951, there was a further rust outbreak, this time spreading throughout the maize-dependent regions of Nigeria and severely damaging crops there. In the Gold Coast the results were even more alarming. Losses to the maize crop drove the Gold Coast price to 89 shillings per hundredweight, an all-time record. The maize shortfall there in 1950 had required the import of 12,300 tons of grain, and the projection for 1951 was for a further 15,000 tons. In June Secretary D. Rhind of the Department of Agricultural and Forestry Research offered a pessimistic forecast in a memorandum circulated by the West African Inter-Territorial Secretariat to the governors of Nigeria, the Gold Coast, Sierra Leone, and Gambia. A storm was brewing:

> Maize rust disease has been known in many parts of the world for a long time, but it has rarely assumed epidemic proportions and generally the damage done is negligible. There are records of previous outbreaks as epidemics in America and West Africa but they appear not to have been as bad as the present epidemic in West Africa. At first it was thought that the West African outbreak in 1950 might be due to unusual weather conditions favouring the fungus, though there was no proof of this theory and not all the facts were in conformity with it. In 1951 the attack has again been extremely severe in Nigeria and the Gold Coast. At present I have no information from other West African territories but they are unlikely to escape.[8]

As a key officer in charge of the colonies' agriculture (and therefore farm economies and food supply) Rhind struggled to find a cause for the rust outbreak and suggested a biological nightmare. He realized that the weather theory was no longer tenable and that the more likely explanation was that the cause was a new and "virulent biological strain of the parasite." American research had already shown at least seven strains of the type of rust that had existed in the New World. The rust fungi were capable of changing their physiological form and "hybridizing, one strain after another, giving rise to different degrees of virulence." Rhind feared that such a new strain had come to stay in West Africa. He went on to express his exasperation with the difficulty of controlling the still mysterious rust:

> There is only one satisfactory way of dealing with this disease and that is to find or breed a variety resistant to it. Removal or burning of old maize plants or infected leaves might control the disease if every single farmer throughout West Africa carried it out, not missing a single scrap of the diseased maize. This is quite impossible. Accidental maize plants occur round habitations or even in the bush and those could carry on the fungus from season to season. The spores by which it is propagated can be blown by wind for immense distances, several hundred miles having been proved possible. Each diseased leaf produces enormous numbers of these spores, which are microscopic in size. Once a focus of the infection starts it can spread at a very rapid rate, miles per day if there is only a gentle breeze to carry the spores. It is therefore quite useless to try to control it by plant sanitation methods. Equally, spraying with fungicides is impracticable. The maize plants will need to be sprayed repeatedly, every fortnight at least, as each new leaf produced by the plant is not spray-protected. . . . The fungus can get actually inside the sheaths of the cob where spray or dusts could not penetrate.[9]

Rhind's memorandum concluded with a proposal to begin a full-scale effort to focus the regional resources of the British colonial administration in West Africa to address the threat to the region's food supply and economic stability.

Over the course of the next two years the effort to control the strange and virulent American rust mobilized the resources of the late colonial structure and reached beyond to include the emerging postwar world of science, multilateral institutions, and American hegemony. Science thus became part of a new postwar world order that recognized the global nature of the rust outbreak itself and put the response to it on a global footing.

The first stage of the colonial West African Inter-Territorial Secretariat Committee's response was to set up a West African research consortium made up of elements of the West African Departments of Agriculture to begin a breeding program to create a rust-resistant maize cultivar. Secretary Rhind proposed a strategy to mobilize the full resources of British colonial science and its international partners:

> The breeding of a resistant variety (which must also be agriculturally satisfactory for such characteristics as yield, life period, palatability, etc.) requires concentrated and continuous research by well-trained and experienced maize breeders and mycologists. It is a problem which is likely to take some time, being limited by the cropping seasons and complexities arising from the maize plants' method of reproduction. It affects all maize-growing areas of West Africa, and may even spread to other parts of the continent. It is therefore a problem particularly well suited for research by a team of scientists working on behalf of all the West African Governments whose results could be expected to find wide application. I therefore consider it a matter of urgent necessity that there be set up, with the least delay possible, a research project to undertake the study of rust disease on maize and the search for resistant maize varieties.[10]

Rhind proposed four possible sites for the research facility: the Ibadan University College of Agriculture (Nigeria), the Kwadaso Central Experimental Station (outside Kumasi in the Gold Coast), the Njala Agricultural Research Station (Sierra Leone), and Moor Plantation (Ibadan, Nigeria). It is noteworthy that after independence each one of these sites became an important institution for national and international agricultural research.

In July 1951 in a joint memorandum from the Committee for Colonial Agricultural Animal Health and Forestry Research the governors of the Gold Coast, Nigeria, Sierra Leone, and Gambia assessed the situation and agreed to fund a research project at Moor Plantation. The governors agreed with Rhind that Moor Plantation's location outside Ibadan was best, since it offered water, electricity, gas, land for test plots, and housing suitable for a resident European (British) research team.[11]

On 18 June 1951 Rhind had also written to S. P. Wiltshire, director of the Commonwealth Mycological Institute at Kew Gardens, broadening the scale of the colonial response and indicating the work done to date:

> The maize rust in West Africa about which you have no doubt heard last year has again appeared in epidemic form in Nigeria and the Gold Coast. I have no news yet from other territories in West Africa about the present position but will be visiting Sierra Leone and Gambia in the next month. The damage which this disease has done and is doing to the present crop is very serious indeed. There are no precise figures of loss of crop but I have seen many fields totally destroyed while less severe loss is inevitable at all places in Nigeria and Gold Coast wherever maize is grown.

The failure to find any West African rust-resistant maize meant that the effort would necessarily have to draw upon a far-flung network of scientific resources. In Nigeria alone, the research team tested forty-five maize varieties and found them all infected, as were all those examined in the Gold Coast.[12]

At about the same time as Rhind was developing his strategy (July 1951), an American agronomist, O. J. Webster, was touring Nigeria on behalf of the Economic Cooperation Agency (ECA), an adjunct program of the Marshall Plan. Webster's presence expanded the international networks further when he suggested that the ECA could help fund a maize breeder from America through the American Point Four program and noted that he had also contacted a maize breeder in Venezuela and another in Mexico.[13]

Despite some hesitation about opening the door to American

technical and institutional involvement in internal British colonial affairs, Geoffrey Herklots at the Colonial Office in London noted the American Webster's idea and proposed his own strategy to contact Dr. M. C. Jenkins at the U.S. Department of Agriculture's Bureau of Plant Industry in Beltsville, Maryland.[14] Herklots held out hope, nonetheless, that British imperial resources would be sufficient to cope with the crisis. His strategy was to avoid involving American experts directly; he nonetheless hoped to obtain U.S. maize varieties for testing at the Nigerian research station and further suggested that a British member of the West African staff be responsible for writing "to other parts of the world where maize is grown as a commercial crop." He also wished to appoint a mycologist from Britain who could, if necessary, visit specialists in the United States. To find the British mycologist, Herklots suggested contacting universities of Cambridge, Birmingham, and Oxford as well as the Rothamsted Experimental Station in Hertfordshire.[15] But despite official reticence to reach beyond British colonial resources, the networks already stretched throughout the British Isles and across the Atlantic. The race of colonial science against nature had begun, and there was no time to lose.

The Global Infrastructure of Science

F. C. Deighton, who in 1949 had been the first scientist to identify the outbreak of the mysterious and virulent rust, was a crop pathologist working for the Sierra Leone Department of Agriculture at its research station at Njala. In September 1949 the uredeospore samples he found on rust-infected maize plants appeared to be the common rust *Puccinia sorghi,* though the spores were unusually large and the damage to the plant far greater than he had seen before. Deighton wondered whether this was a new variety or mutation. He had dutifully passed on a rust-infected leaf cutting to Kew Gardens. In October 1951 the important news reached the Colonial Office in London (via West Africa) that mycologist G. R. Bisby at Kew had examined West African rust spore specimens and determined by examining the heretofore missing samples of the dormant teliospore phase that the West African rust was not a variant of the common *P. sorghi,* but *P. polysora,* a type of rust never before reported outside the Americas; for West Africa an unwelcome visitor, indeed. Moreover, retrospective study of earlier samples of rust on West African maize confirmed that the new fungus had not been present in the Gold Coast or Sierra Leone before Deighton's report of 1949. Transatlantic science then expressed itself again when the American G. B. Cummins at Purdue University in Indiana confirmed Bisby's research at Kew. Cummins had been the first to identify the *P. polysora* rust in Alabama in 1941.[16]

The race had begun, but the rust was already far ahead of its pursuers. Spring 1952 reports from field agents at the Mugugu agricultural research station in Kenya of a rust outbreak along the Kenya coast and in the Taveta highlands confirmed that the new *P. polysora* American rust had appeared there as well. The minor incident reported first in Sierra Leone now took on continentwide implications well beyond British colonial borders. From February through March 1953, reports of new outbreaks came in from Katanga (Belgian Congo), Bas Congo, Nyasaland, Yagambi (Congo), and Réunion Island near Madagascar.

Colonial science was now in hot pursuit. By the 1953 crop year, an expanding international network of individuals and institutions were actively engaged in the search for a solution. That network included the newly formed West Africa Maize Rust Unit at Moor Plantation, Nigeria, the East African Agricultural and Forestry Research Organization, the U.S. State Department's Economic Cooperation Agency, and several institutions in Britain, especially the Commonwealth Mycological Institute at Kew Gardens and the London School of Hygiene and Tropical Medicine. In March 1953, W. R. Stanton, director of Research at the West Africa Maize Rust Unit, received Colonial Office funds to travel to Lake Como, Italy, to attend the International Congress of Genetics and to visit the

Italian Istituto di Cerealicultura in Bergamo. In his application for travel funding he foresaw his trip as an ideal opportunity to consult with the wider scientific community about the American rust in Africa. This visit and Stanton's contacts expanded the professional networks and for the first time engaged international geneticists in the effort to deal with the rust threat that had heretofore involved only plant breeders and fungus specialists. In the course of 1953 the network broadened to include the New World and the multilateral agencies of the brave new postwar world: the Food and Agriculture Organization, the Iowa Tropical Research Center in Antigua, Purdue University, the U.S. embassies in Rome and in Caracas, Venezuela, and the Rockefeller Foundation.[17] In effect, the multilateral reconstruction of postwar Europe had global implications that affected African colonial networks of agricultural research. Communications between these groups, the British Colonial Office, and even French colonial officials were surprisingly seamless. In the case of the Rockefeller Foundation, its European offices were located, strangely enough, at the Colonial Office in London itself.

From early on in the fight against American rust (after October 1951 known to be *P. polysora*), colonial crop scientists and mycologists had agreed that the best strategy for combating the disease was to breed on test plots a resistant variety of maize suitable to West African conditions. To do this required two sets of activities. First, researchers had to find maize varieties that resisted the rust, provided acceptable crop yields, and tolerated tropical field conditions. Very early on in the process they concluded that no West African maize types had any resistance whatsoever.[18] Therefore, the scientists' search for genetic varieties for crossbreeding would have to call upon a massive international effort to assemble genetic materials from biomes all over the world, and especially the New World, where *P. polysora* had long existed in a balance with maize. It was a sign of the times that none of the parties involved proposed a farm-level survey of fields in West Africa or called on the local knowledge of African farmers.

Second, for crop scientists to develop a resistant variety would take many successive crop cycles, including multiple test plots cultivated in a given year. The problem was that the *P. polysora* rust reached its infectious uredeospore stage only once a year, and then it receded into its dormant teliospore stage. Infecting a test plot only once a season would be far too slow a process to yield the results as quickly as they were needed. How could researchers artificially infect test plots many times within a single season to identify resistant cultivars and to breed resistance? In September 1953 J. M. Waterston, research adviser from the Moor Plantation Maize Rust Research Unit, announced a breakthrough, a payoff for the investment in the colonial science infrastructure. The Moor Plantation research team had developed a method of preserving *P. polysora* uredeospores in vitro, thus allowing field scientists and breeders to induce infection of many different maize varieties on multiple test plots for rust resistance over the course of a calendar year.[19] The primary concern now became the need to assemble as much genetic diversity as possible to develop resistance in a maize variety that would also provide viable yields in West Africa's humid and semihumid conditions. In August 1953 early results at Moor Plantation on twenty-nine assay plots to compare eight Mexican types and one South African variety had shown promising results.

Not surprisingly, lowland Mexican varieties showed the most promising resistance to the rust that had its origins in that New World ecology.[20] But the resistance of the South African maize type Tsolo (brought to West Africa in 1941 from South Africa's dry eastern Cape region) was especially encouraging, even though the variety's yield was less well adapted to West Africa's humid zones than many of the Mexican types.

Even as West African efforts proceeded, East Africa had become infected. In June 1952 reports of maize rust outbreaks from East Africa (Tanganyika, southern Kenya, and coastal Kenya) reached West African agricultural officers. In late 1952, research efforts intensified, as rust infections, confirmed to be *P. polysora*, appeared elsewhere in East Africa and spurred the development of a parallel research effort by the East African Agricultural and Forestry Research Organization at its research station at Mugugu in Central Province, Kenya. On the East African side the results were as discouraging as they had been early on in West Africa. None of the East African varieties showed any resistance whatsoever to the new rust.

With research under way on two sides of the continent the international networks of late colonial empire came into play, fostering the fledgling links between colonial structure and postwar multilateral agencies emerging in Europe and across the Atlantic. Global science expressed its newly forged geographic range and undertook the task of transporting plant genetic materials across global biomes to retrace the steps of the previous expansion of maize and to uncover the path of *P. polysora* itself. As early as December 1950, and proceeding apace in 1951, 1952, and 1953, varieties of maize had begun arriving in West Africa from Ontario (Canada), Michigan, Maryland, and North Carolina in the United States, India, Ceylon, Malaya, Mexico, the Caribbean, Venezuela, and South Africa.[21] From around the globe small shipments of seed corn, a few ounces at a time, arrived at Moor Plantation by post and courier from private seed companies, university faculty, embassies, research stations, and farmers' associations. Once the research stations began formal trials (after 1953) the trickle of genetic materials to West and East Africa research stations became a flood. By May 1953 Harold H. Storey, secretary for East African Agricultural Research, claimed that multilateral sources and his personal contacts in Central America had provided more than two hundred sample seed types. Institutional sources for these maize varieties included

The West African Maize Rust Research Unit, Ibadan (Nigeria)
The Agricultural Experimental Station, Medellín (Colombia)
Purdue University School of Agriculture (Indiana, USA)
The Imperial College of Tropical Agriculture, Trinidad

The Rockefeller Foundation (New York, USA)
The Department of Agriculture, Salisbury (Southern Rhodesia)
The Director of Agriculture, Zomba (Nyasaland)
The Department of Agriculture, Pretoria (South Africa)

The process of transferring and exchanging genetic materials between continents and across global biomes appeared effortless and quick, and rationally planned, in contrast to the first biological transfers of the Columbian exchange, which had been halting and often haphazard. Scientists' transfer of rust-resistant genes was nonetheless tardy, given that *P. polysora* had already moved much more quickly than their belated efforts. Nonetheless, late colonial global science was on the case.

Denouement

After the 1953 crop season the enigmatic American rust receded as quickly as it had appeared on the scene three years earlier. A rust-resistant maize cultivar produced by global science networks was ready for release in 1957, but colonial African agricultural authorities never in fact made it available. We are left with the question of how to account for the virulence of the original outbreak of American rust and its sudden subsidence.

Nobel laureate Norman Borlaug speculates that a long-dormant "fossil gene" from West African maize's New World ancestry emerged among local African maize varieties to reassert resistance to the fungus, as a result of African farmers' mass selection of maize genotypes that had survived the rust. According to Borlaug's hypothesis, American rust has origins in the same New World ecology that spawned its host crop, maize. Two species of rust exist as parasites on maize in Latin America and the Caribbean, *Puccinia sorghi* and *Puccinia polysora*. The former predominates at higher altitudes and lower temperatures and the latter at lower elevations and higher temperatures. One or the other of these rusts infects virtually every maize plant throughout the natural range in Mexico, Central America, the Caribbean, and northern South America. In areas of the New World where both rust types exist, local maize types have evolved resistance to both. Therefore, because of a biotic balance between host and parasite, damage to New World maize has been rare.[22] The rapid expansion of maize around the globe in the last half-millennium and its more recent commercialization as colonial food supply, however, apparently disrupted this balance and set the stage for emergence in the second half of the twentieth century of American rust—the aggressive fungus that had somehow been left behind in the original Columbian exchange.

At the time maize was first imported to West Africa in the sixteenth century, *Puccinia sorghi* (now called common rust) traveled with it, and West African maize had coexisted in a balance between the host and rust from the beginning. Examples of *P. sorghi* rust existing benignly on African maize had been reported for decades, and pathologists had assumed it to have been present for much longer, presumably since the initial arrival of maize. The second rust, *Puccinia polysora,* however, appears to have remained behind in the New World until the mid-twentieth century. Around 1949, for some still unknown reason, it encountered African maizes. Over the course of four hundred generations of maize and ten thousand generations of the pathogen, West African maize varieties had lost their resistance.[23] Once unleashed on Old World tropics, American rust attacked in Old World tropical and subtropical maize landscapes that had no resistance and where humidity and temperature conditions stimulated the rust's life cycle of infection, dormancy, and transfer.

Even if the infectious American rust aecidospores could not travel by wind across expanses of ocean, they could nonetheless exist on live host tissues that traveled by sea and air, probably in the form of food shipments (maize transported either as grain or still in the husk) that went from the New World to colonial possessions in tropical Africa during and after World War II.[24] Sierra Leone was just such a trans-shipment point for Allied food supplies from the Americas. Spores that arrived in Sierra Leone thus found a tropical climate and host maize that had long ago lost resistance to it. It apparently took three years for resistant genes to emerge in West African maize or for the rust itself to lose its aggressive nature.

Borlaug also speculates, with little direct evidence to back him up but with laudable populist sentiment, that African farmers themselves directed the emergence of resistant strains, by selecting seed from surviving plants that displayed evidence of resistance.[25] Confirming Borlaug's speculation would require extensive field studies among the older generation of West African farmers, a worthwhile project but one well beyond the scope of this chapter.

The third possibility is simply ecological serendipity: fungi—or insects, for that matter—that assert themselves with great virulence at certain times and in certain places often recede inexplicably. Environmental history is not linear, but more often conjunctural, resulting from a critical mass of factors and their interaction at a specific place and time. This is the case with the groundnut *(Senecio vulgaris)* fungus *P. lageruphorae* and with the Sri Lankan "brown bug," a case in which the conjuncture of local conditions and ecologies seems to have brought disease and plants into balance on their own terms.[26] Science, and farmers as well, seem to accept such anomalies, even if they cannot fully explain them.

Finally, American rust has more recently taken a new name and a new persona. Now known as Southern rust, *P. polysora* is no longer an unwelcome stranger, but an endemic risk, a familiar foe, especially in tropical maize fields, where its sporadic reappearance is unpredictable and almost quixotic. It is one of many threats to Africa's food supply that both farmers and scientists face from season to season and for which maize breeders constantly seek resistant genes.

It is especially notable that in the archival paper trail at the British Public Records Office, at Kew Gardens, and in scientific professional publications on the *P. polysora* outbreak African farmers are voiceless in the scientific policy narrative, both as victims and as agronomic actors. There is no recorded evidence of their response to the crisis or mention of any concern among colonial scientists to use farmers' knowledge as a potential source of insight into the disease. This neglect is surprising in the face of evidence that New World and West African farmers spent many generations adapting New World maize germplasms into forms that fit the ecological and economic niches in African landscapes.

The lack of interest in or attention to local knowledge and conditions marked a new era in colonial science. In the 1920s and 1930s, and into the 1950s, there existed a solid tradition of colonial ecologists whose work reflected a deep understanding of African practice of agriculture, husbandry, and environmental management. This was especially evident in the work of people like Colin Trapnell, William Allen, and John Ford whose thinking emerged from extensive and pragmatic field studies in an older tradition of ecological study.[27] This tradition declined in the late 1930s and 1940s, superseded by the emergence of "experts" trained more formally in the sciences of agriculture and medicine. This formal training, at sites like the West Indies School of Tropical Agriculture in Trinidad, emphasized formal research and the use of international scientific networks. The new trend manifested itself in the programs of forced terracing, anti-trypanosomiasis campaigns, and livestock management schemes.[28] The new paradigm ignored the local in favor of the overarching structures of professional science.

Internationally, however, *P. polysora* continued to expand its territorial range, appearing here and there in Africa and continuing eastward into Asia and the Pacific. British colonial records for maize rust in West Africa end abruptly after 1953, though Kew Gardens' mycological program continued to record minor *P. polysora* outbreaks around the globe into the 1970s and to collect maize leaf tissue samples. In June 2000 Brian Spooner, head mycologist at Kew Gardens, took me to the basement of the mycology laboratory. He pulled from the shelves a large binder marked *Puccinia* and left me to ponder it. Pasted to the first page was a small, withered scrap of corn leaf—Deighton's original cutting sent from Sierra Leone in September 1949.

7

Breeding SR-52: The Politics of Science and Race in Southern Africa

In late 1960 something remarkable happened in the history of African maize, seemingly almost miraculously. Scientists at a little-known research station in the small and politically fragile Federation of Rhodesia and Nyasaland in central Africa announced the release of a new kind of maize seed. They had given it the unimaginative name SR-52 (that is, Southern Rhodesia 52).[1] This SR-52 was a hybrid maize, a type of seed that heretofore had emerged only from the modern science of powerful and well-financed private seed companies in the United States, then as now the dominant world power in agricultural research on maize. As it happened, SR-52 was an anomaly, a "single-cross" hybrid, of the sort that most crop specialists and economists of the time claimed could never succeed. It would be, they said, too expensive to produce, too susceptible to genetic contamination, and too much in need of kid-glove treatment from farmers in Africa too inexperienced to manage it. But if Hickory King was the Colt .45 that had won southern Africa at the opening of the twentieth century, SR-52 was the Apollo 11 moon landing, for it ushered African maize into a brave new world of scientific breeding. To coin another metaphor, SR-52 was a thoroughbred, the Seabiscuit that won the race and then sired another generation of winners.[2]

Indeed, SR-52 was a miracle of sorts, one that transformed African landscapes, racial politics, and diets over the next forty years. In its first two decades after its release in 1960, SR-52 raised yields on African commercial farms by more than 300 percent over the previous decade. In 1950 only 22 percent of the large-scale farms in the region had planted hybrid maize; by 1967, more than 93 percent were doing so, and it was virtually all SR-52.[3] Moreover, the putatively sensitive and quirky SR-52 had real staying power: by its fiftieth birthday (in 2010), it and its descendants will still appear in maize fields all over central and southern Africa and as far west as Cameroon. Its progeny and that of its inbred parent lines still prosper in drought-prone zones and sandy soils; over its life span SR-52 has provided both mealies porridge for working-class tables and baby corn for upscale international markets.

Quite a tale lies behind SR-52's extraordinary success, one about politics, climate, the phenomenon of colonial science, land hunger, dedicated and innovative scientists, and the genetics of maize. The story takes place not just in Southern Rhodesia (Zimbabwe, after 1980), but also in the rest of the abortive federation, in Zambia and in Malawi, whose history with maize resulted in those three nations' beginning the twenty-first century among the world's top ten consumers of maize as human food. The peculiar resourceful character of SR-52 itself provides the leitmotif for the history of African maize in the second half of the twentieth century.

SR-52's Political Crèche

On 1 August 1953 an act of the British Parliament gave birth to the Federation of Rhodesia and Nyasaland, the dream of the generation of white farmers and entrepreneurs who had succeeded Cecil Rhodes's 1890 pioneer column of white settlers, and a nightmare for Africans, who looked forward to independence from European hegemony. Advocates for the amalgamation of the British colonial possessions of central Africa (a self-governing colony, Southern Rhodesia, and two protectorates, Northern Rhodesia and

Nyasaland) saw the federation as an opportunity to set up a larger internal market—to introduce economy of scale—and increase regional control for the white settlers of Southern Rhodesia, who sought greater autonomy from London and Labour Party meddling. Those goals were precisely what such African nationalists as the firebrands Kenneth Kaunda, Joshua Nkomo, and Hastings Banda feared, as they sought to consolidate postwar economic growth by forming a political movement for independence and control over the area's nascent urban areas and considerable resources in mining, agriculture, migrant labor.

Commercial farms in Southern Rhodesia, and to some extent in Northern Rhodesia as well, offered an economic and political base for the population of white settlers and an economic engine fueled by their control of the best-watered, most salubrious areas. At least one version of the settlement story is that black farmers before 1890 had occupied lighter, sandy soils in that low and middle veld that suited tilling by hand hoe. Settlers, however, saw the promise of the heavy red basaltic soils that held great potential for maize but were heavily wooded and expensive to clear. The settlers argued that the wood derived from the clearing of forest to establish maize plots served as a fuel for the curing of tobacco, which proved to be Southern Rhodesia's major export crop up to 1965. While there is some agronomic evidence for this view, it is a loaded take on local history that smacks of the old South African empty-lands myth.[4] In the end, however, black smallholder farms ended up on lands with scantier and less predictable rainfall, while large white-owned commercial farms claimed land that was ideal for large-scale commercial cultivation of maize and cash crops.

Settlers' economic plans included prospects for an export trade in tobacco and cotton, as well as regional markets for food crops to feed the growing towns and mining communities of the copper belt and southern Africa as a whole. Maize was obviously a crop that could supply food for local markets, but Rhodesian farmers also expected to compete on the London Corn Exchange with American and Argentine maize—in the international export markets.

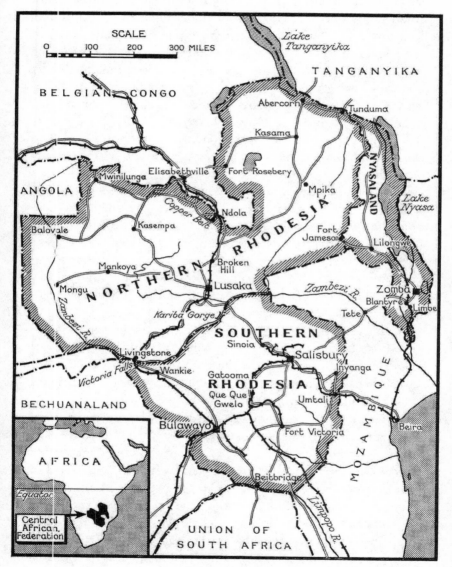

25. The Federation of Rhodesia and Nyasaland, 1954–1963.

In the early 1950s the federation's white population was still small (seven thousand in Nyasaland, ten times that in Northern Rhodesia, but perhaps two hundred thousand in Southern Rhodesia). In the first five years of the federation some hundred thousand white settlers came to take up land and jobs in its burgeoning economy, and in 1955 the federation government unilaterally decided to construct the Kariba Dam on the Zambezi, behind Victoria Falls, at the cost of almost one hundred million pounds sterling.[5] Prices for Northern Rhodesia's copper soared in the early 1950s and buoyed up the entire economy. Southern Rhodesia had the beginnings of an industrial base and a vibrant urban center at Salisbury. If Nyasaland was the poor cousin of the federation, a sleepy colony of small farms—predominantly black—it had potential wealth in the form of tobacco and cotton. The federation's economic growth rate in its early years was remarkable, surpassing that of the postwar economies of Europe and the United States.[6] Those were heady times, and Southern Rhodesia's white settler population under the leadership of Roy Welensky and Godfrey Huggins (later Lord Malvern) was firmly in charge of the new federation's political mechanisms.

From its earliest foundation in 1890, the government of Southern Rhodesia (until 1923 under control of the British South Africa Company) was committed to the economic health of white-run agriculture; white settlers were its political and social backbone. White commercial farms thrived on black farm labor, primarily from Malawi and Mozambique, a strong frontier entrepreneurial spirit, and the grit and determination of settler families.[7] Annual agricultural shows in the capital at Salisbury rewarded innovation in crops and techniques. Particular settler families specialized in their favorite type of "flat white" maize, imported from South African relatives or directly from the United States.[8] Farmers took great pride in displaying their locally adapted white maize types at the "Cob Showing" at the annual Salisbury Agricultural Fair. Over the course of two generations, Southern Rhodesian farms developed a series of disease-resistant white dent maizes that were well suited to the particular conditions of the well-watered highveld.[9]

Well before the federation came into being, indeed as early as 1897, Southern Rhodesia's Department of Agriculture began to focus almost entirely on securing the future of this frontier offshoot of European agriculture and its access to good land, markets, and research support. From 1909 on, the Botanical Experimental Station (later renamed the Salisbury Agricultural Research Station) commenced testing crops with economic potential. In 1930 Dan McLaughlin, a maize researcher at the Salisbury Agricultural Research Station, found in a South African journal an article reprinted from an American publication that detailed work going on in the United States on maize hybrids. He passed it on to a fellow plant breeder, Harry (H. C.) Arnold, who was inspired by the possibilities for southern Africa. In 1932 Arnold, then director of the Salisbury Agricultural Research Station, began a special effort to improve the varieties of open-pollinated maize being grown on commercial farms, by inbreeding classic white dent to form distinct "heterotic" groups, the first step toward hybridization. He had in mind the maize varieties already cultivated by the white families on the Rhodesian commercial farms. Those open-pollinated varieties had local and exotic names like Salisbury White, Hickory King, Louisiana Hickory, Southern Cross, and Potchefstroom Pearl.[10] The effort seems largely to have ignored the colorful local flint types that African farmers cultivated as a vegetable on their intercropped farms. Over the next three decades Arnold assembled a team of focused, well-paid maize breeders that included Alan Rattray, S. W. Nelson, Rob Olver, Rex Tattersfield, and Mike Caulfield, who fought doggedly for government support, and got it.

As in the United States during the same period, these researchers worked to isolate inbred lines from locally adapted open-pollinated varieties gathered from the region and from U.S. sources. It was soon obvious that the local flat white dent types, rather than the imported varieties, were the most promising, since local farmers had already selected strains resistant to ear rot and maize streak virus, diseases common in the rainy highveld. At the outset of the hybrid experiments, Harry Arnold had selected fifty ears of Salisbury White and begun a systematic process of inbreeding for yield and resistance to disease. By the late 1940s he had produced an ideal parent line, called, rather clinically, N3.[11]

This research effort in Southern Rhodesia, however, took place in a somewhat different context than did U.S. work on hybrids, for colonial legislation set up a distinctive political climate for agricultural research. Rhodesia's 1930 Land Apportionment Act had established the Department of Agriculture and Lands, which was to assume responsibility for agricultural research and whose mission was openly framed as working exclusively for the needs of "European" agriculture. African agriculture, by contrast, fell to the Department of Native Affairs and, later, the Department of Native Development, which was specifically concerned with such issues as cattle destocking and changing African agronomic practice, rather than exploring African smallholders' crop needs.[12] The colonial government was intent on the view that research on new crops was to benefit European farmers and to be suited to the kind of land they occupied and the intense rural entrepreneurship of the predominantly Scottish and Irish immigrant farmers. The group tested the first experimental hybrids as early as 1938, not long after such work had begun in the U.S. private sector under such legendary names in agricultural research and politics as Henry Wallace of Pioneer Hi-Bred Seed Company.

Hybrid maize became the priority for research in Southern Rhodesia because of its potential in the prime heavy soil of agricultural lands controlled by white commercial farmers and its possibilities for supplying the growing urban centers with the white maize flour that had rapidly become the regional staple; as mentioned earlier, production of white maize for flour dramatically overshadowed production of the hard-starch flinty maize of African household gardens and intercropped farm fields. Maize in the minds of Southern Rhodesia's planners and plant breeders was decidedly an industrial grain, and not a vegetable to round out rural diets.

In the 1930s, however, the success of white commercial agriculture was not assured; in 1932 the government sought to protect the white farm economy from depressed agricultural prices by launching a new structure, the Maize Control Board, which controlled who could buy maize in bulk and what prices would prevail for which types of farmers (black and white). The goal was partially to stabilize the supply for the cities and mines, but more particularly to prop up prices to sustain white farms. In fact, in virtually all African settler colonies during the economic slump of the 1930s white commercial farms felt threatened by the fact that African farmers could produce maize surpluses more cheaply than could settler farms, which had invested heavily in farm infrastructure like dams, mechanization of processing, and use of inputs. By the mid-1930s each of the governments of Southern Rhodesia, Northern Rhodesia, and Kenya had set up its own version of a maize control act. (In Kenya, it was the 1935 Native Produce Ordinance.) These marketing acts had similar features that created extra-market advantages for settler commercial farms, by

1. Creating official government buying stations in favored European farming areas, but no parallel structure in African areas
2. Enforcing a dual pricing scheme, with higher maize prices for settler farms
3. Establishing restrictions on grain movement from African areas to towns and mines (in some areas white farms' proximity to rail lines and roads had already effectively accomplished that).[13]

African farmers were not banned from growing maize, as is sometimes alleged, but the Maize Control Board created clear obstacles to their participation in the market on equal terms.[14]

For the Southern Rhodesian case, agricultural economists Melinda Smale and Thomas Jayne have plotted the per capita grain production of African producers in the period from 1914 to 1994. Their findings show a significant drop from 1935 until the Maize Control Board was dissolved in the early 1940s to stimulate war production.[15] A decided downward trend is obvious throughout the colonial period. Smale and Jayne make three additional telling points. First, they argue that the cost of subsidizing for settler-grown maize was paid, in effect, by African farmers and consumers, rather than by colonial governments, which made such payments fiscally and politically sustainable during that era. Second, the growth in large-scale commercial agriculture that these payments underwrote helped sustain political and financial support for hybrid-maize research in the postwar years. Finally, the introduction of hammer mills in rural towns in the 1920s to produce maize flour for town dwellers gave an enormous advantage to the commercial farmers who produced white dent maize. The more traditional African grains sorghum and millet, required dehulling before milling and Africans' local flint maizes were neither uniform in color nor suitable for either the local hammer mills or the later industrial steel roller mills.[16]

In 1935 the Northern Rhodesian government and the colonial government in Kenya, another settler colony, also introduced maize marketing controls, though in Northern Rhodesia the fixed prices served to stimulate African production in key areas like the Tonga Plateau, rather than to undercut African maize farmers, as happened in Southern Rhodesia.[17]

During World War II, when central Africa's white farmers and researchers alike served in North Africa and Italy, maize production and prices soared to meet wartime demand, but little work on hybrids took place. But in 1948 the Cambridge-educated Alan Rattray, one of the Arnold team, took on the post of officer in charge of the Salisbury Agricultural Research Station. Rattray gradually shifted the station's scientific emphasis from general agronomy to plant breeding; he set up the station's fields for ideal conditions of soil fertility, slope, and tillage for plant breeding and even coldheartedly sent away the group of retired mules kept on station to provide manure (chemical fertilizer was in vogue after the war, and the mules had therefore become redundant). Rattray was, by all accounts, all business.[18]

In Southern Rhodesia, government support for maize research was not just a commitment to agricultural modernization; it was part of a larger plan to ensure the economic and social base of white rule. Though the maize breeders largely kept their noses to the grindstone and buried in their research station test plots, the emphasis on the study of improved and hybrid maize took place at precisely the same time as the systematic attempts both before and during the federation to undermine African maize production and secure the valuable agricultural lands for the settler economy.

But the story was not so simple. The Salisbury Agricultural Research Station team released its first hybrid in the 1949–1950 season, a double-cross called SR-1, and followed that with a series of other double-cross types through the 1950s. Yes, Rattray and Arnold's researchers had certainly sought to maximize yield for the white commercial farms that dominated the region's high plateau, but they were also aware of the potential for drought in central Africa, so different from the U.S. corn belt and from the well-watered Rhodesian plateau. Though the primary mission was to prop up large commercial farms, the hybrid research offered potential advantages to black farmers in drought-prone areas as well. When rainfall 250 millimeters below normal fell on their test plots, the yields of experimental hybrids were 68 percent higher than yields for the usual open-pollinated types, such as the old standby Hickory King still sown on small African farms. The hybrid advantage in years of poor rainfall was even greater than the 50 percent advantage over nonhybrids that such experimental types enjoyed in a normal year.[19] These results in the 1946–1947 season reinforced and justified the research emphasis on hybrid seed as the proper strategy for Southern Rhodesia. It remains unclear whether these were unintended consequences or a foreshadowing of the potential shown by research on hybrids to address broader needs once the nation was under black majority rule. The focus on research continued apace, and in 1948 the government created a new post, that of director of the Department of Research and Special Services at the Salisbury Agricultural Research Station, to coordinate and support agricultural research. Two years later the government further extended its research investments by grouping conservation and extension services in a separate unit, thus leaving the team of maize scientists to focus exclusively on their hybrid breeding mission.[20]

By the time the federation came into being in 1953, the maize-breeding research in the tiny, landlocked settler colony was a world-class effort, single-mindedly directed toward a specific goal: to develop a maize type suited to the defined needs of the particular ecology of Rhodesia's highland plateau. The dominant role of Southern Rhodesia in the federation had an immediate impact: the federation government concentrated its agricultural research efforts at the Salisbury Agricultural Research Station, while efforts elsewhere in the federation atrophied, including parallel efforts that had begun in Northern Rhodesia (such as the Mazabuka Experimental Gardens, established in 1913). Commercial farmers and Southern Rhodesia's needs were front and center in the research effort.[21]

Though scientists at the Salisbury Agricultural Research Station had begun working with double-cross hybrids in the late 1930s and 1940s, by the late 1950s they had taken the unusual step (especially by contrast with their American counterparts) of concentrating their efforts on single-cross techniques. Single-cross hybrids are the product of two parent lines—or heterotic groups—that breeders select and "clean" over several generations to focus on particularly desirable traits. After establishing the parent lines in a pure form, breeders then cross-pollinate the two parent lines to achieve heterosis, the effect of "hybrid vigor." Such "single-crosses" often produce high yields in the first generation (F-1), but generally the yield deteriorates rapidly in the next (F-2) generation, unless farmers buy new seeds each season. The second-generation decline in yield with single-crosses is even more pronounced than with double-cross hybrids. Moreover, the parent lines need to be kept pure, as breeders must turn the process of commercial seed production over to farmers, who need to manage the parent lines on a large scale to multiply the seed and release it commercially for use on farms. This is an expensive and risky process, which usually discourages using single-cross hybrids for commercial purposes. Most hybrids until the late 1950s had been the more stable double-cross types.

By 1958, however, the research team had two clean parent lines in place. One was the Salisbury White dent that Arnold had begun in 1932, when the idea of a hybrid effort had first surfaced. The other was from Alan Rattray's own selection of fifty ears of another flat white dent, Southern Cross, which had a slightly higher yield than Salisbury White but a later maturity. Rattray had begun introgressing that parent line soon after he returned from the war, in 1945. In October 1958 researchers crossed the two lines on test plots on the grounds of the Salisbury Agricultural Research Station. At harvest in 1959 the results were so astonishing that team member Rex Tattersfield weighed the crop twice, to be certain he had not mistakenly combined two plots' yield. Thus SR-52 was born, and the team rushed word of the blessed event to the farmers' Seed Coop to urge immediate release of the seed to farmers, an unprecedented recommendation from the usually cautious Rattray-Arnold team.[22] In almost no time, SR-52 was released to commercial farmers for the 1960 planting season.

The Salisbury researchers had taken a risk in pursuing a single-cross strategy, but they had the benefit of working with the wide range of locally adapted, open-pollinated white dents already ideally suited to the burgeoning markets for maize flour. The dents were uniformly white, and the soft starch of the grains' endosperm matched the needs of the local industrial mills, which supplied regional markets and had already superseded household hand-grinding.[23]

The world's first commercially successful single-cross hybrid, SR-52 spread across the high-potential lands of central Africa like a firestorm. This result predated Asia's Green Revolution by half a decade and highlighted the economic benefits of investment in research. In the first years after its 1960 release SR-52 produced yields 46 percent higher than its parent, Southern Cross, which to that point had been the most popular type among white commercial farmers. By comparison with the decade before 1960, SR-52 provided tripled maize yields, an average of almost five tons per hectare. By the years 1976 to 1980, just before Zimbabwe gained independence, virtually all commercial farms adopted it. On some farms SR-52's results even surpassed U.S. corn belt yields. Farmers in the federation and Rhodesia who realized yields of one hundred bags per hectare, a world-class achievement, joined the Rhodesia's elite Ten-Ton Club.

Because it required early planting and a long maturation period, SR-52 rewarded intense management from farmers: the use of tractors, nitrogen fertilizer, and large-scale plots of farmland made Rhodesia's commercial farms viable on the world market. Irrigation in October before the rains on many Rhodesian commercial farms gave germination a head start and increased the prodigious yields even more. Moreover, by the late 1960s Rhodesia was producing its own nitrogen fertilizer (ammonium nitrate), thereby giving the crop the potential to become a major export commodity.

The planting of SR-52 boosted farmers' production almost exponentially. But SR-52 seed was also an exportable good in demand all over southern Africa. To satisfy the burgeoning regional markets, the Department of Agriculture and the Seed Maize Association had established a sophisticated network of farms to produce SR-52 seed, a complex and demanding task. In fact, some might claim that the decentralized network of seed-producing farms was as defining an achievement as the development of the SR-52 plant itself, given the tricky nature of single-cross hybrids.[24] South Africa proved to be a major customer, as were the then Portuguese colonies of Angola and Mozambique. Shaba Province in the young Congo state was potentially the best customer for SR-52, but the Congo crisis of the early 1960s forestalled that possibility.

Within its genetic makeup and political context, however, SR-52 had tragic flaws worthy of a Euripidean hero. Even though it richly rewarded the farmer's careful management of plant population and fertilizer, SR-52 did not like sharing the farmer's attention with other crops. As a monocrop, it was not happy being intercropped, according to the usual practice with maize on African smallholder farms. These characteristics were built into both its genetic makeup and the narrow institutional view of central Africa's future. Like most of the highest-yielding hybrids, SR-52 required a long growing season (157 days at 1,500 meters of elevation) and therefore had to be planted early—mid-October, in central African terms. For each day's delay in that early planting there is a 1 to 2 percent loss in yield. As a consequence, SR-52 manifested a strong bias against hand tillage and richly rewarded farmers who had the cap-

26. Garmany's roller mill, Salisbury, ca. 1920.

ital or credit to prepare their fields with tractors or a span of oxen in October. That is well before the soil is friable enough to prepare by hand with family labor, as black farmers would have had to do.

Without easy access to credit, input delivery, cash reserves, or well-watered land, planting SR-52 could be a risky business. Moreover, the advantages of SR-52 as a single-cross hybrid depended on farmers' ability and willingness to repurchase new seed each season or risk a substantial loss in yield. Research done in Southern Rhodesia in the late 1960s showed that "recycled" SR-52 lost 46 percent of its yield in the second generation, while a more conventional double-cross (SR-12) lost only 19 percent.[25] Additional risks arose in drought-prone areas, where the rains often cease early. Drought, by causing the hermaphroditic SR-52's male parts (the pollen-bearing tassels) to emerge later than its female parts (the silk), could lower its yield. In sum, the miracle crop was ideally suited to white commercial farms and poorly adapted to Southern Rhodesia's small black farms, particularly under white Rhodesian rule in the period before 1980, when the deck was stacked against black farmers. And in the decades after 1960, political unrest began to surface, as black leaders emerged who envisioned a noncolonial future.

The Federation of Rhodesia and Nyasaland collapsed in 1963 as a result of its political failure to cope with the "winds of change" locally and internationally, and especially its inability to overcome the concerted opposition of the African leadership to subordination to the priorities of the white settlers.[26] Zambia's and Malawi's independence followed in 1964. In 1965, the white settler government of Southern Rhodesia sought to forestall black majority rule in the former colony through its Unilateral Declaration of Independence (UDI), which resulted in international trade sanctions against Rhodesia's major export crops, especially tobacco, and industrial goods, like steel. Because maize was traded within the region, it offered a way to avoid the economic sting of sanctions from Europe and the Commonwealth. Southern Rhodesia's ability by the late 1960s to produce its own nitrogen fertilizer to sustain its commercial maize production was also decisive in setting the national course. Maize exports, including sales of SR-52 seed, were a major part of the efforts to sidestep sanctions.

How could the Rhodesian state cope with international pressure and economic isolation? Here the research capacity and investments in human capital of the federation period and before paid off. Tobacco, the cash crop most affected by the post-1965 sanctions, did not share the same landscape as SR-52. In fact, commercial farmers planted their tobacco, their most profitable crop, on sandy soil and in zones with marginal rainfall. Besides tobacco, maize and cotton were the only two crops that could hope to replace the income lost to the sanctions.[27]

To counteract the losses from export-oriented cash crops, the Southern Rhodesian research effort redirected its energies in the late 1960s toward developing the short-stature, short-season (early-maturing) triple-cross hybrids R200, R201, and R215 (now bearing the prefix 'R' for "Rhodesia"), which were adapted to the sandy soils in areas of low rainfall and intended to allow white tobacco farmers to reinvest in maize.[28] Of these, the early-maturing short-season crop R201 was especially well suited to the task.

But there were interesting unintended consequences as well. In 1968 Mike Caulfield, the maize breeder responsible for monitoring seed sales to commercial farms, checked his sales projection chart for the new hybrid R201, the most promising seed developed for the sandy soils and lower rainfall on tobacco farms. He noticed a substantial gap between the projected needs of his target farmers and the actual sales recorded in his hand-drawn graph. He wondered about the source of the unexpected demand and speculated that it was, in fact, black smallholders who had observed the value to them of the early-maturing hybrid R201 and had begun to buy it in substantial quantities. Clearly, black farmers had been aware of the hybrids' advantages and were ready to invest in new varieties as soon as early-maturing seed became available. This response, in fact, anticipated by several years what economist Carl Eicher was later to call Zimbabwe's second maize revolution. Despite the lack of any specific research focus on black farmers, smallholders had initiated a significant and unanticipated response to the rise of hybrid-based agriculture.[29] The by then venerable SR-52 played a role in this change as well, since a parent line of R201 was N3, one of the two parent lines derived from Rattray's original inbred Southern Cross.

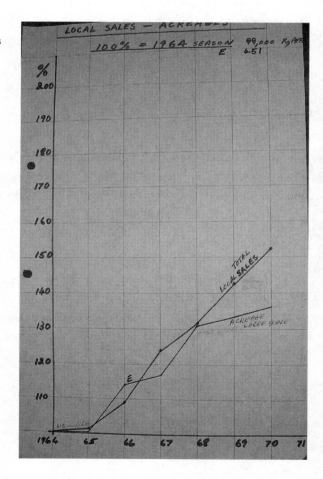

27.
Mike Caulfield's original seed sales chart, 1968.

Thus, when UDI Rhodesia succumbed to the forces of black resistance and international pressure in 1980, to become majority-
rule Zimbabwe, its talented and well-paid maize breeders had on
hand hybrid-maize types that at least ostensibly suited a variety of
conditions, including those which black farmers faced on the so-
called Communal Lands and Resettlement Areas. The result, in
the period between 1980 and 1990, was a transformation that included large, high-potential commercial farms, small-scale commercial farms, and also black farms on marginal Communal
Lands, where sandy soils and the specter of drought challenged
farmers' ability to participate in the SR-52 revolution.

Zimbabwe's Second Maize Revolution, 1980–1986

The negotiated end in 1980 to Rhodesia / Zimbabwe's guerilla war
brought about a rapid change in the social and political landscape;
on the economic side it left largely intact the structure of large commercial farms on the most productive lands and black smallholder
farms on what the new government called Communal Lands. The
new majority-rule government also established a smaller number
of farms for ex-guerillas in Resettlement Areas. Whereas virtually
all the commercial farms producing maize and tobacco were in areas of high rainfall, more than 90 percent of the Communal Lands
and Resettlement Areas were in areas considered marginal for
maize production.[30] Nevertheless, in the period from 1980 to 1986
maize yield on African smallholder farms doubled and total production increased more than sevenfold. By the end of the first decade of majority rule in Zimbabwe, virtually 100 percent of Zimbabwe's maize fields were planted with the hybrids developed in the
1960s, including the short-season triple-crosses suited to drought-
prone areas.[31]

The scale of these changes after 1980, however, and particularly
African smallholders' wholesale shift to hybrid seeds, was an indication of more than just the success of the hybrid seed program.

Other factors included the end of the guerilla war and the return of security to rural areas, heavy investment in rural roads and infrastructure, and the increasing availability of credit to small farms. Success, however, brought vulnerability. Many black farms verged on a maize monoculture and depended heavily on good rains and government credit, expectations that were not always realized.[32]

Quite clearly, there was a peace dividend as black farmers gained access to land, credit, and the kinds of hybrid seed (the R200 series) developed during the sanctions period of the late 1960s. In fact, it may have been black farmers' use of the R200 series of short-cycle hybrids that allowed them to fare better during drought than the commercial farms that continued to use the long-cycle, drought-vulnerable SR-52. The steady increase in smallholders' share in national maize production throughout this period also perhaps suggests a leveling of incomes in rural areas. This evidence of increased use of hybrids and expanding participation by small farms in producing the national food supply has caused a number of economists and planners to wax ebullient about prospects for Africa's Green Revolution, led by hybrid maize and investments in research.[33]

The national figures from the postindependence period, however, are somewhat deceptive. First, average yields on large-scale commercial farms (5.0 tons per hectare) were and are still far above the average yield on Communal Lands (1.5 tons per hectare). And the results, which were not evenly distributed among black smallholders, fueled prosperity among those on the best land but resulted in land hunger and the sense of a promise unfulfilled among others. The bulk of the smallholder production increases after 1980 came from only 15 to 20 percent of the Communal Land farmers, especially those in the three Mashonaland districts—Zimbabwe's Maize Belt—where soil and rainfall conditions are optimal. These three well-watered areas account for only a quarter of the smallholder maize-farming area, but by the mid-1990s those lands supplied three-quarters of the maize sold.[34] In fact, black smallholders overall sell only about half their maize, meanwhile consuming the other half at the household level. Maize accounts for about half the average Zimbabwean's total calories, a fact that suggests sizable differences in economic class among Zimbabwe's farmers, which has fueled continuing land hunger and provided fodder for populist politicians.

Aggregate figures suggest that Zimbabwe's maize-based agrarian economy is quite vulnerable to drought, as was evident in the years 1983–1984, 1984–1985, and especially in 1987–1988 and 2002–2003. Is it possible that the research that helped make maize an economic engine for the national economy in the period from 1960 to 1990 might also help address the vulnerability to drought?

The history of maize in southern Africa is more than the saga of Southern Rhodesia / Zimbabwe. Elsewhere in the region, contrasting human dramas also reveal much about the distinctive but overlapping historical experiences of farms and rural economies. Maize has been one of the actors in those dramas, through which we can trace the fortunes of Southern Rhodesia's (junior) partners in the abortive federation (1953–1963)—Northern Rhodesia (Zambia) and Nyasaland (Malawi)—and in Kenya, Africa's other settler colony.

Zambia: Reinventing SR-52

The collapse of the federation in 1964 pushed the two Rhodesias onto diametrically opposed paths: black majority rule for Zambia in the north and a white settler oligarchy in Rhodesia in the south. In 1964 Kenneth Kaunda brought an independent Zambia onto the stage as a "frontline" state sworn to oppose both apartheid in South Africa and white minority rule in Rhodesia. As Africa's most urbanized new nation, Zambia faced the task of building a rural economy that could feed its mining-based labor force and also be an engine of growth unto itself. Though committed to majority rule, Kenneth Kaunda's government sought to build upon an economy in which the large-scale farms resembled those of Southern Rhodesia. Many of Zambia's small African farms suffered from the absence of male labor, after the exodus to the mines and cities; nonetheless, a vibrant rural population on the Tonga Plateau near national markets had long since learned that maize was both a crop to feed the household and one that could be sold on national markets to great advantage. On the Tonga Plateau the arrival of the moldboard iron plow early in the twentieth century had already begun to transform gender roles and the cropping mix on small farms.[35] Historian Kenneth Vickery quotes ecologists Colin Trapnell and J. H. Clothier, who remarked on these distinctive changes on the Tonga Plateau's black farmers in the late 1930s, when large white commercial farms were still struggling:

> Increasingly large acreages are plowed for maize as a commercial crop. The women still maintain the domestic garden of some two or three acres of mixed kaffir corn [sorghum] and maize, with traditional subsidiary crops, but in the large maize gardens made by men there is a general tendency to abandon traditional crops and to adopt changes in cultivation . . . commonly the Kaffir corn area [of the wives' gardens] has become the adjunct of a large maize garden from which the surplus will be sold to the neighboring store.[36]

In this transition of gendered labor and shift to a market orientation the older flint maize types cultivated and consumed as a vegetable crop had given way to white dent types cultivated as grains, especially Hickory King and Salisbury White, to fit the needs of urban mills and urban consumers.

For the market-oriented of the Tonga plateau, maize proved an attractive crop. As Vickery put it:

> The important point is that maize could be eaten or sold. This was a great advantage. Production of surplus maize created not only purchasing power but a more secure food supply as well. The Pla-

teau Tonga farmer did not have to make a difficult choice between producing a cash crop for the market and producing food for the household, or try to balance labor between them.[37]

By 1935 the Northern Rhodesian settler-farmers had successfully lobbied the colonial government for the protection of the Maize Control Ordinance, akin to the act that Southern Rhodesia had introduced and the ordinance that white settlers in Kenya had lobbied to achieve that year. In Northern Rhodesia, however, the effect was somewhat different than the one colonial administrators had anticipated (suppression of black African participation in the maize trade). In 1936 colonial authorities anticipated that black Africans would contribute 50,000 bags of maize to the market; in fact, those farmers marketed 234,000 bags, with 75 percent of that coming from the Tonga Plateau. The guaranteed price from the board had been an incentive to increase African production, since it allowed black farmers to do an end run around traders and get a higher overall price than in the past. This situation was quite different from that in Southern Rhodesia, where the board action had lowered prices for African producers who lost their bargaining power and their incentives. Nevertheless, guaranteed prices subsidized by the state kept large-scale white farms viable through the 1930s, and the need for wartime food production created sufficient demand to relax the pricing structures.

But Tonga was an anomaly. In other parts of Northern Rhodesia, demands for mine labor slowed agricultural change, or at least the adoption of maize as a grain crop. In the Bemba region, for example, Kenneth Kaunda related that during his youth, in the 1920s, 1930s, and up to World War II, maize remained largely a garden crop supplementing millet and sorghum. Flint maizes were widespread in kitchen gardens. Farm households consumed maize fresh, but ate millet as the preferred staple cereal. It was only in the 1940s that maize as a food and as a crop achieved a dominating role.[38] The 3:1 price ratio for maize set in 1935 by the Maize Control

Board and intended to support white farm economies ended in 1942 as the wartime colonial government sought to stimulate over-all production.

Changes in the late 1940s and early 1950s proceeded apace, as large white-owned farms along the rail lines continued producing white dent maize types for national markets. In 1949 to 1953, large-scale farms doubled their yields, from 1.3 to 2.7 tons per hectare, as the demand for maize also increased twofold in that period. As the white settler population grew in Northern Rhodesia in the postwar period, the acreage owned by settlers of European extraction doubled, but so did native African production. By the early fifties, Africans produced 45 to 50 percent of all maize marketed in Northern Rhodesia.[39]

When the federation period ended in 1964, farms north of the Zambezi had already embarked on the hybrid-maize revolution and had begun the mechanization that made expansion of settlement possible. By the early 1950s virtually all settlers' farms were mechanized, and from the early 1960s until 1980, Zambia's maize farmers had increased production by 400 percent, expanding onto new land, raising yields, and tapping into heavy government subsidies for fertilizers. But the changes were not distributed evenly. While large-scale white-controlled farms adopted SR-52 wholeheartedly, only about 30 percent of the small-scale farms planted it. As mentioned earlier, SR-52 was a tough taskmaster, in that it insisted on early planting, heavy weeding at prescribed times, nitrogen fertilizer, and the use of tractors, or at least ox-drawn steel plows. Farmers who could not fulfill those demands had to settle for planting a mix of older white dents like Hickory King or mixing them with the older flinty types that farmers kept on hand or traded among themselves.

Unlike Rhodesia's UDI government, however, Kenneth Kaunda's single-party government invested heavily in making small-scale farming work out and viewed SR-52 not as a product of settler research, but as an endowment that would or could transform Zambian agriculture as a whole.[40] Though raised on a mission station in northern Zambia and experienced as a teacher, political organizer, and small-scale trader, Kenneth Kaunda harbored a passion for farming. He viewed agriculture as a critical contributor to independent Zambia's economic foundations, not merely as a means of sustaining a cheap food supply.[41] Unlike the turn toward state intervention in agriculture taken by Julius Nyerere in socialist Tanzania through its Ujamaa experiment, Kaunda's policy in Zambia was to invest systematically in smallholder agriculture. In the late federation period and early in the post-1964 independence years, Zambia's large-scale European-owned farms along the railway lines had accepted SR-52 at a rate similar to that of their counterparts in Southern Rhodesia. The strategy of Kaunda's ruling United National Independence Party, however, was to reduce its dependence on such farms and encourage the participation of the small farmers in remote areas who had formed an important part of Kaunda's political constituency. But only about 30 percent of Zambia's small-scale farmers in the late 1960s had begun to plant SR-52, because its requirements fit poorly with those farmers' seasonal calendar, their need to plant other crops first (cassava, millet, and so on), and distance from transport and markets.[42]

President Kaunda's fledgling government attempted to overcome these obstacles by providing substantial subsidies for fertilizer inputs and offering "panterritorial" pricing that in effect compensated for the higher transportation costs in remote areas. The National Agricultural Marketing Board had come 180 degrees in reversing the old colonial Maize Control Board's price fixing to benefit European farmers. Kaunda's system aimed to stimulate the small-scale farmers to produce supplies of maize for urban residents, for whom white maize meal had now become the staple food. The small rural households and urban workers that benefited had, in fact, formed the UNIP's primary political constituency; they now received a tangible reward for their support.

In the wake of independence, when Zambia lost its own agricultural research institutions to Salisbury, Zambian participation in the hybrid revolution began to unravel. At the breakup of the federation, the Rhodesian government had given the new Zambian government (and Malawi as well) eight pounds of SR-52 parent seed. But in the mid-1970s, in the attempt to build its own seed program for SR-52, train Zambians as maize breeders, and establish new priorities for breeding, a disaster took place. The local maize seed producers contaminated both parent lines of Zambia's SR-52.[43] Neither Zambia nor Malawi, in fact, had the capacity to produce SR-52 seed commercially, since they did not have the decentralized seed farm structure that had sustained Rhodesia's economy. Since well over half of Zambia's maize came from farms smaller than five hectares, between the late 1960s and the mid-1970s, national maize production foundered and nationally hectarage planted to maize was stagnant. Zambia's national economy thus suffered simultaneously from its insistence on maintaining sanctions against Rhodesia and the beginnings of the collapse of the global copper prices that had sustained its investments in small-scale agriculture.

In the late 1970s international donors from Scandinavia, the United Nations system, and the United States renewed their interest in Zambia's maize research efforts and helped the government reconstitute the moribund Mazabuka Maize Research Institute. The institute's first task was to rebuild its own stocks of SR-52 (despite the sanctions, Rhodesian breeders apparently helped in this effort, a tribute to their commitment to science).[44] In 1984 the Mazabuka Maize Research Institute released MM752, its new version of SR-52, boasting that it had a yield 20 percent higher than the original SR-52. Between 1984 and 1992 Zambia's maize breeders, warming to their task, released nine new white dent hybrids adapted to Zambia's local conditions and two new early-maturing flints intended for small farm gardens. Yields for these hybrids were more than half again as much as those for the local types they replaced.[45] Zambian farmers now had a significantly wider range of choices, allowing them to respond to their own farm strategies, place in the market, and judgments about climate and soil.

The new varieties had significant advantages. Thanks to the plants' resistance to disease and maturation up to seven weeks earlier than SR-52's, farmers could plant late and still get better yields than with older types, even without fertilizer. Finally, these were double- and triple-cross hybrids that were cheaper to reproduce than the single-cross SR-52, and farmers could replant them for a year or two and not lose yield dramatically. The national investment strategy worked pretty well: use of hybrid maize on small farms doubled, to 60 percent of total maize-growing area, and nationally maize production rose by 137 percent over the fifteen years from 1975 to 1990. Small farms in Zambia at that point produced 80 percent of the country's maize.[46] Kenneth Kaunda's agrarian strategy for attaining equity and an alternative model to Rhodesia had had marvelous results, or so it seemed.

Zambia had underwritten its investments in marketing, maize research, and inputs such as fertilizer by tapping into booming world market prices for copper during the early 1970s, sustained in part by the Vietnam War and the Cold War in general. Copper prices crashed in the wake of the 1973 oil crisis and failed to recover as some had predicted. Meanwhile, Zambia's fertilizer subsidies, which had so greatly benefited small-scale maize farmers, reached a whopping 17 percent of the total national budget by 1988. Zambia's progressive policy of subsidizing small farmers in areas remote from towns and transportation quickly became the target of the new liberalization ideologies sweeping the International Monetary Fund and the World Bank. The organizations that had generously pushed loans on Zambia in the 1970s now sought economic reform and made Zambia's maize system a target. Urban maize price increases fostered urban unrest in 1986 and 1990, and President Kaunda's maize policy was a major campaign issue in the 1991 election that brought down his UNIP government. A sporadic liberalization process followed, interrupted by the impact of regional drought. The contraction of credit and the loss of price supports on remote small-scale farms by the mid-1990s caused a reduction of more than 15 percent in land planted to hybrid maize. Many farms reverted to older, less market-oriented crops like sorghum, millet, groundnuts, and beans. At the opening of the twenty-first century, however, Zambia's agrarian economy seems on the brink of returning to a dual structure, in which the role of maize is unclear.

Malawi: Maize Is Our Life

Malawi ranks number three (behind Lesotho and Zambia) on the most recent list of the countries with the highest per capita consumption of maize in the world. More than half (54 percent) the total calories in Malawi's national diet come from maize. Yet unlike in Zimbabwe and Zambia, in Malawi the maize comes from fields that are planted with many other crops. Monocropping characterized most of the production of South African, Lesotho, and Rhodesian farms. Even though maize occupies 80 percent of all cultivated land in Malawi, 70 percent of all maize plots contain between two and five other crops as well.[47] Malawi, in this and in other ways, offers an alternative version of the way maize has spread into the southern African landscape. Paradoxically, Malawi's agriculture, seemingly conservative in its acceptance of hybrids, has its basis almost entirely in New World plants. The historical interaction of Malawi's farmers with these plants shows a fascinating dynamism.

Pauline Peters, a social anthropologist, spent a decade (1986–1997) interviewing farmers and observing farm practices in one agroecological zone, Malawi's Shire highlands, a densely settled southern region where small farms struggle to produce on intensely cultivated plots that average about one and a half hectares.[48] Her 1990 survey showed that about a third of total household income came from maize produced and consumed at home, while less than

a fifth came from the sale of food crops; 11 percent came from tobacco sales, and about a twelfth from the sale of small livestock.
The remaining third came from various sorts of "off-farm" income. In these circumstances, the primary farm strategy for balancing income sources, including crops, and food supply is to diversify
as a way of managing the unpredictable nature of climate, ecology,
market prices, and political circumstances.

Despite the fact that maize is the primary product of their fields,
Malawians in the Shire highlands mix their maize plantings with
pigeon peas, beans, pumpkins, cowpeas, and groundnuts. They
plant maize and the intercropped plants in holes adjacent to each
other along a ridge. The growing maize stalks then provide a kind
of trellising for vines and shade for germinating seeds. The mix of
maize, cowpeas, and pumpkins is an especially old formula, dating
back at least to the nineteenth century, and thus when farmers
add groundnuts to the mix, the fields, interestingly, contain all New
World plants. To those crops the farmers have also added cassava,
another New World plant, whose cuttings they dig in along the
edges of maize fields.[49] The farms that plant burley tobacco (another New World plant and Malawi's largest export crop) have
done so by alternating it annually with maize.

A central issue for those who have studied Malawi's production
of maize has been the slow pace at which Malawi's farmers have
accepted the new hybrids that have been so readily accepted by
farms in Zambia, Zimbabwe, and South Africa. In Peters's 1986
survey only 2 percent of the sample households were growing hybrid varieties; the vast majority of farmers preferred to sow their
local varieties—mainly hard-starched flints—which many referred
to as *chimanga ndi moyo* (the maize of the ancestors). From a late
twentieth-century perspective this agricultural system seems to
have been slow to respond to new markets and possibilities. Yet the
system's adaptation to new crops and ways of introducing plants
into local ecologies demonstrates a remarkable historical dynamism, akin to West Africa's invention of forest fallow. In this case,

short-cycle flint maizes played a systemic role over time in a quite dynamic agricultural landscape.

Peters identifies three fundamental reasons Malawi's farmers were so reluctant to give up their older maize types.[50] First, the new dent hybrids had a soft endosperm readily damaged by weevils in storage; the harder starch of the older local flint types stored much longer. Second, Malawi's small-scale farmers had a low income by comparison with those in neighboring countries, and it was difficult to purchase new hybrid seeds and fertilizer every year. Third, the soft endosperm of the dent hybrids did not pound well—the bran did not separate easily from the flour—and women asserted to researchers that the taste was unsatisfactory. In response to agricultural surveys, Malawian women overwhelmingly preferred the storable "flintiness" of the maize of the ancestors—the types derived from the original maize brought from the New World. The farmers thus chose storability, taste, and texture above yield. Moreover, even richer farmers who adopted new dent hybrids because of their impressive yield continued to sow some land to the older, early-maturing flints, which they used as in-kind food payments to workers they hired to harvest their hybrids.[51]

By the early 1990s Malawi was quickly catching up to its neighbors in hybrid plantings. By the time of Peters's survey in 1990, half of her sample farms were growing hybrid maize, including new derivatives of the classic SR-52 and R201. The key ingredients in the change directly addressed the issues listed earlier. The Ministry of Agriculture had shifted its focus away from Malawi's few large-scale commercial farms, which had cared less about the storage of the crops that they intended to market quickly than about maximizing their yield. The new policy made hybrid seed freely available to small farms. But most important, the government supported Malawi's professional maize breeders in developing new, "flintier" hybrids (called variously semiflints and semidents) that had harder starches, tasted like the old flints, and still retained the early-maturing qualities that made the corn available before dent hybrids and often allowed the plants to "escape" droughts, by maturing before the rains failed. Many of these new semiflints, in fact, derived from Rattray's N3, perhaps the world's most prolific inbred parent line. Southern Cross genes still roam central Africa.[52]

Malawi's farmers have thus made the transition from old varieties to "maize modern" without abandoning the smallholder strategies of intercropping, which allow them to balance the need to intensify their labor on small farms, the need for involvement in the market, and the need to diversify both their income and household food supplies. Peters summarizes the farmers' rationale for their practice of what others have seen as "one of the great glories of African science" and "a skilled craft."[53] First, intercropping of grains and legumes protects the soil fertility of land in permanent use (and where crop rotation is less feasible); second, the diversity of crops on a particular plot reduces the risk of drought, pests, and crop disease; third, it allows what Peters insightfully calls "sequential decision making"—that is, responding to the exact progression of conditions in any given year. Finally, intercropping is a labor-intensive way of economizing on land, while preserving the household food supply.[54] If Malawi's smallholders produce less maize than their neighbors do, the farmers have nonetheless advanced their craft to the point where they can withstand the vagaries of both economy and climate.

Malawi's case contrasts markedly with the mineral rich economies of the Rhodesias (later Zambia and Zimbabwe) that joined it in the ill-fated federation. With Malawi's primarily agricultural base, the small African farms were less affected by massive labor migration to the mines and colonial cities than was the case in either of the Rhodesias. Malawi also had a much smaller and less politically ambitious European population than did other parts of the federation. After independence its economy was primarily agricultural and small in scale. Maize was a food crop first and foremost, rather than a cash crop linking the farm to a national cash economy. The previously told story of Malawi's larger neighbor Zimbabwe offers a useful contrast. If SR-52 was Southern Rhodesia's marquee success story, then the saga of the humble but venerable flinty maize has been Malawi's distinctive tale.

Kenya: The Political Ecology of Colonialism

Like the federation, Kenya had the economy and political structure of a colony, where white commercial farmers exerted influence over the majority black rural population and sought to take a line independent from that of the Colonial Office in London. Lacking the mineral resources of southern Africa, however, Kenya had a more purely agrarian base, one that included an active and growing group of middle-class black farmers in the 1950s.[55] Maize in the period before 1900 had been one of several cereal crops produced on small farms around Lake Victoria and in the central highlands, but nowhere a primary focus of farmers. Yellow maizes were the dominant type until the early twentieth century, when settlers experimented with white dents from southern Africa such as Hickory King, which offered higher yields than the older intercropped types.[56] By the 1920s maize had become the dominant cereal on both African-run and European-settled farms. The biggest jump had come during World War I, when a famine had forced farmers to eat their millet seed, which government programs and small-scale merchants then replaced with maize in following seasons. Kenyan farmers never looked back.[57]

As maize became the dominant staple in the interwar years, the most prevalent form came to be a white dent resulting from settlers' selections, a type that they came to call the Kenya Flat White complex. The Kenya Flat White complex served much the same role as the Salisbury White, Southern Cross, and Hickory King varieties in Rhodesia. The beginnings of hybrid research in Kenya came later (1955) than in Southern Rhodesia and also had a different focus. In late colonial Kenya, agricultural policy, influenced by the Mau Mau emergencies, had begun to focus on the needs

of small African farms in the central highlands and the food supply for the growing urban centers. In 1958 a survey of Nairobi
households had shown that maize accounted for 80 percent of all
starchy-staple calories, a dependence second only to that of Lusaka
(Northern Rhodesia).[58] Sustaining maize production was thus both
an agricultural and a political issue.

Kenya's first maize research program began in 1955 (that is to
say, two years after the end of the Mau Mau emergency) in Kitale,
in the Central Highlands. In the first years, Kenya's maize breeders
achieved only poor results with their attempts to cross Kenya Flat
maize with improved white dent types from South Africa, Australia, Rhodesia, and the U.S. corn belt. In 1959 the Rockefeller Foundation funded a trip by a colonial Kenyan maize breeder named
M. N. Harrison to Mexico and Colombia, where he collected more
than a hundred local unimproved maize types from equatorial
highland areas similar to those in East Africa. Over the next few
years the Kitale research program tried 124 crosses in its search for
a hybrid "hit."

The breakthrough finally came in the 1961 field trials on the
eve of Kenyan independence, when the researchers crossed their
open-pollinated Kitale Synthetic II with an obscure, unimproved
Ecuadoran land race bearing the mundane title Ecuador 573. The
result of this strange mixed marriage, as with SR-52's genesis, was
nevertheless magical. Kenya's H611 had a yield 40 percent better
than had the older nonhybrid Kitale Synthetic II, and as a varietal
hybrid it had a low production cost. The Kenya government officially released Kitale H611 in 1964, just after Kenya gained independence. Kitale H611 fitted not only the ecology of central Kenya,
but also its political climate, since its characteristics were ideally
suited both to large farms and to the many new African small-
holdings that had proliferated in Kenya in the late colonial period.
The breakthrough offered Kenyan farmers a choice between the
new hybrid, ideally suited to well-capitalized farms, and the older
but still productive Kitale Synthetic, which matched the needs of

smaller or part-time farmers. Kitale H611's propagation among Kenyan farmers in the post-1965 period was even faster than the spread of hybrid maize among U.S. corn belt farmers in the 1930s and 1940s.[59] This rapid spread certainly resulted partly from the qualities of the seed, but it also benefited from an elaborate network of roads, extension agents, and small African and Asian shops that distributed it. Strangely, Kenya's planners had modeled their marketing of hybrid maize on the successful campaign to popularize Wilkinson Sword razor blades to rural customers: "Every stockist [shopkeeper] an extension agent" was the mantra. And it worked.

The success of the Kitale program was followed by similar success in a later, parallel crop research program in Katumane, near Machakos in the drier zones to the east. Just as Kitale H611 had

28. "Every stockist an extension agent," Cargill Seeds, Mbulo, Tanzania, 1996.

spread across the highlands, Katumane maize allowed farms in short-season or drought-prone areas to participate in the national maize "madness." As a result, Kenya was able to keep pace with its neighbors to the south. In 1960, Kenya's small farms planted 44 percent of their land in maize; a decade later it was 51.4 percent. In 2002 maize accounted for about half the calories consumed by Kenyans, showing remarkable continuity, given the country's economic tribulations in the last three decades of the twentieth century.

Kenya's case contrasts in interesting ways with the southern African experience. Late colonial policy in Kenya had focused its maize research to benefit African-run farms, more as Zambia did in the post-independence period, and Zimbabwe after 1980. By the end of the twentieth century a snapshot of maize planting in different countries indicates a construction of "maize modern" that reflects their individual histories. In 1990, for example, Kenyans planted 62 percent of their maize with hybrid seed, while in 1999 the figure was 85 percent. In Malawi it was only 11 percent in 1990 but 39 percent in 1999. In the same decade, Zambia's hybrid percentage actually declined from 72 to 62 and Zimbabwe's dropped from 96 to 91 percent.[60]

This snapshot, however, conveys the impression of a trajectory toward more hybrids and national markets, a central role for agricultural science in the late colonial period, and independence in the last half of the twentieth century. Or are the changes episodic steps rather than a progressive march forward? General circumstances affecting the political ecology of maize make future directions less than clear.

Maize and Malaria

8

In late summer 1998 a severe and deadly malaria epidemic broke out in northwest Ethiopia. In the single district of Burie alone there were more than 42,000 cases and more than 740 deaths—in other words, 47 percent of the district's population reported having the disease over the period from June to December. But the outbreak was not evenly distributed. In one locality the local health official closed ninety houses, the residents dead of a deadly falciparum malaria. In a local school in another location half the children lost one or both parents. Yet another location reported few cases, if any at all. Some places had no cases in September but hundreds in November. The patchwork quilt of infection brought confusion and desperation. Farmers blamed *zar* spirits, consulted diviners, and sought propitiation; others sought succor from the few government health stations scattered around the district.

The spotty "hot zone" outbreak continued through the fall months and gradually dissipated by January 1999. What made this outbreak puzzling was that it affected a region that had never known epidemic malaria before. Indeed, in the mid-1970s it had had neither mosquitoes nor malaria.[1]

What accounted for this new infection, its devastating intensity, and its uneven distribution across the rural landscape? No immediate answer to that question came from either residents or regional health officials. Some residents recalled unusually intense indoor mosquito-biting activity that August and September. Rural Orthodox Christian folk in some districts had blamed the *zar*, an evil spirit, and tried to placate it with sacrifices of oxen and sheep. Others called it *tesibo* (typhus) and avoided contact with their neighbors for fear of infection. Few if any local residents linked the deadly outbreak with *nidad* or *enqetqet*, two names for malaria, a disease that they associated with distant lowland areas and not their highland plateaus.[2]

New research, however, raises a novel question: whether some relation existed between the unwonted intensity and geographic distribution of malaria cases in northwest Ethiopia and the new agroecology of maize production that affected the area at the same time. Addressing this question has required the detective skills of an environmental historian, an epidemiologist from the Ministry of Health, and a young entomologist who shared a deep interest in both the human tragedy and the environmental puzzle.[3] Part of the solution goes back to a hunch pursued by Yemane Ye-ebiyo, a young Ethiopian entomologist engaged in a doctoral program at the Harvard School of Public Health about the same time the epidemic was taking place in northwest Ethiopia.

Yemane's idea came from his curiosity about the feeding habits of the larva of *Anopheles arabiensis,* the mosquito that is the main vector for malaria in Ethiopia. He was specifically interested in what effect food supply had on the survival and growth of anopheles larvae and how successful the larvae were in advancing to the pupa stage of their life cycle. Since larval food supply included bits of edible materials present in their breeding habitat (usually small muddy catchments around human settlements) he set up a controlled field experiment at Zway, in southern Ethiopia, to see the effect of plant pollen on the anopheles breeding cycle. For his experiment he chose maize, the most commonly grown domestic plant around Zway and increasingly Ethiopia's major food crop. His hypothesis was that since maize is one of Africa's few wind-pollinated (as opposed to self-pollinating or insect-pollinated) field crops, maize pollen might indeed fall on the turbid water near homesteads that is the anopheles' favored breeding site. What, he asked, was the value of maize as a food source for the mosquito larva?

Yemane's experiments compared breeding sites that contained maize pollen and those which did not. Furthermore, he compared sites that were within ten meters of a maize field with sites that were at least fifty meters from a field; overall, he sought to compare the effect of maize pollen on the number of larvae that survived to the pupa stage, and thus to adulthood, the speed of this growth, and the size of the adults that emerged from the breeding sites.

His results were astonishing. Of the larvae that had maize pollen as a nearby food source virtually all (94.1 percent) advanced to the pupa stage, while only 13 percent of those located at more than fifty meters' distance survived to the pupa stage. In a parallel experiment Yemane compared breeding sites sprinkled daily with maize pollen with those from which maize pollen was excluded. The result was that maize pollen–fed mosquito larvae developed into pupae more than ten times more often than did larvae that were not so nourished.[4] What was even more telling was that the maize-fed adults were almost uniformly 13 percent larger than those deprived of maize pollen. Earlier work had already established that the larger the adult, the longer the life span, and thus the greater the possibility of a dense mosquito population with more opportunities to carry parasites from one human victim to another.[5] This point could be explained as a part of what an entomologist would call the extrinsic life cycle of the parasite, a cycle that had to be complete before the female anopheles bite could carry the disease. In sum, Yemane's controlled experiments yielded three major criteria that allowed speculation about a maize-malaria link:

1. Maize fields (pollen sources) had to exist in close proximity to breeding sites (within ten meters to be critical). What was the distance between maize field and house?

2. More rapid and numerous development of larvae to the pupa stage resulted in higher survival rates among the adults and a denser adult population. Was a denser concentration of mosquitoes evident at the site of the epidemic?

3. Adult mosquitoes that as larvae have fed on maize pollen are larger, longer lived, and able to transmit infection more effectively, factors that increase their *vectorial capacity* (and therefore bite rate). But was a reservoir of human parasites also available from which the anopheles could draw their infection?

Was maize an integral part of a human-mosquito landscape that fostered malaria? And could a maize-malaria linkage be proved in an actual epidemic situation?

With these data in mind, I approached Andrew Spielman, professor at the Harvard School of Public Health and coauthor of Yemane's paper, about the possibility of testing those findings in an actual agroecological setting, in this case northwest Ethiopia's Burie district, an area at the heart of the 1998 epidemic where I had lived for two years in the mid-1970s.[6] The Harvard team then put me in contact with Asnakew Kebede, an epidemiologist with the Ethiopian Ministry of Health working in the West Gojjam zone that included Burie. After comparing our field experience and disciplinary perspectives, Asnakaw and I resolved to test the hypothesis that the intensity and spatial dimensions of the 1998 epidemic were a product of several coincidental factors, including the new agroecology of maize, that contributed to a disruption of a long-standing disease equilibrium that had held malaria in check at the research site until the recent outbreak.

In May 2002 we traveled to Burie and began to collect case data from local government health centers and clinics as well as agricultural data from the local offices of the Ministry of Agriculture. We had also gathered statistics and qualitative information on agriculture and environment from various national research institutes on crop production and surveyed local landscapes to identify patterns of cropping and human settlement. We collected observations and bits of local oral tradition from farmers affected by both the malaria epidemic and the emergent agroecology of maize. As we traveled to the malaria-stricken agricultural areas outside the town of Burie, we inspected farmsteads for mosquito breeding sites, proximity to sources of maize pollen, and humans that served as parasite reservoirs.[7]

Collection and assessment of climate data presented particular difficulties, for the National Meteorological Service had closed its Burie data collection station some years before. We therefore approached farm unit managers at the adjacent privatized Bir Sheleko state farm, who graciously provided data from their farm, whose boundary runs parallel to the border with Burie, five kilometers to the east. These data provided us with proxy information about Burie's daily temperature and rainfall for the period from 1997 to 2000.[8]

Burie's historical highland setting is a subset of the classic ox plow agroecology in the Ethiopian highlands. Historically, the farming system has taken shape within a bimodal climate regime (summer wet, winter dry) and annual cropping of endemic cereals (teff and eleusine) as well as cereals (wheat and barley) that fit into Ethiopia's niche as one of the world centers of secondary diffusion (wheat and barley), pulses (chickpeas, horsebeans, field peas), and oil seeds (nug, safflower). As in other areas of Ethiopia, in Gojjam farmers fine-tune their cultivation calendar and agronomic practices to soil types and annual variation in climate (especially rainfall, the key limiting factor). They relied on their carefully adjusted crop rotation of cereals and nitrogen-fixing pulses to maintain soil fertility and to hedge their chances with the variable climate. Anthropologist Allan Hoben, in his pathbreaking fieldwork on land tenure in the area, described in great detail the elaborate schemes by which local farmers claimed scattered plots that balanced soil types and locations, to hedge subsistence strategies.

Burie's agroecology and its relation to malaria had evolved over many generations of smallholder agriculture and the transition in Burie's regional role in trade. Gazing out at the agrarian landscapes that surrounded the town of Burie in the 1970s, one could easily witness the persistent practices that gave its fields their characteristic appearance. Except for a few small market towns along the road, dispersed farm homesteads marked the rural landscape, scattered across the patchwork quilt of cultivated plots and pasture. Farmers chose their crops according to the character of the soil: teff for black soils, barley, wheat, or a bit of maize for red clay soils, *dagusa* (finger millet) for mixed brown soils. Because each farmer desired a mix of cereals and pulses as a hedge against drought or crop disease, each household claimed the use of several plots of varying elevation, soil type, and proximity to the homestead.[9] Highland farmers, exclaiming, "Maret endimert!" ("As the land chooses!"), claimed that their crop mix was determined by the plot soils, rather than the soil adjusted to suit the crop. There was also a balancing of field crops (cereals and pulses) and garden crops (horticultural crops—vegetables, herbs, and spices). The latter sort, carefully fertilized with manure and ash, and tended by women, thus constituted, both spatially and socially, a distinct agrarian space.

The end result was the patchwork landscape of small plots growing diverse crops. From Hoben's classic work and my own observations in the mid-1970s I had in my mind's eye a human landscape dotted by homesteads, each surrounded by a fenced garden and fields. This image hardly seemed to support one of the criteria for a maize-malaria link—the proximity of maize fields to breeding sites near the dwelling. I suspected that the maize-malaria hypothesis from Yemane's work on pollen and larvae would founder on that issue alone. I was wrong about that.

Burie's farm settings range from midaltitude plateau, around 2,000–2,300 meters in elevation, to a marginal lower area descending southward into the Abbay (Blue Nile) gorge, where population is sparse, with settlements occurring mainly along the new road

29. Highland rural compound, 1844.

that crosses the Abbay River. The road retraces the old caravan route that linked coffee-rich Wallaga to Burie, continuing thence to the old imperial capital at Gondar and points north. Truck and bus transport on the road, which was completed in 1988, has been a central feature of the Burie region's recent and growing involvement in national grain markets. The switch to motorized transport may have been a factor too, though goods and people have long traveled along that ancient route.

Burie is generally representative of Gojjam agroecology as a whole, in that its farming and cropping system followed the pattern of using upland red clay soils for certain cereals (barley, wheat, eleusine) and bottom black vertisols for teff, a crop that is tolerant of waterlogging. Burie farmers also planted pulses, whether in relay cropping with teff or on their own, on the of bottomlands that retained residual moisture after the rains. Like most microecologies

30. Highland rural compound, 1975.

31. Highland domestic compound: new maize cultivation, 2002.

in the highlands, Burie also exhibited distinctive local variations on the ox plow theme. Farmers in the Burie area preserved leguminous, nitrogen-fixing *Croton macrostachys* trees in their fields that maintained soil fertility and provided shade for livestock.[10] This farming system has, with slight variations, sustained itself for the past two millennia, a model of stability, and one that national and international agricultural specialists consider to offer high potential. Moreover, West Gojjam was one of the few areas of Ethiopia where consistent rainfall would support new types of long-maturing but very high-yielding hybrid maize.

Burie farmers used little chemical fertilizer before 1980, though they had long applied manure and ash on household garden crops such as maize, kale, capsicum, and herbs cultivated inside their fenced homestead compounds, called *gwaro maret* (household garden plot). In this setting, maize plants were local, early-maturing varieties that germinated with the first rains, tasseled in the cool months of June and July, and then produced green ears in August and September. Roasted or boiled, these *eshet* (young ears, in Amharic) could quell hunger pangs when food shortages preceded the main harvests in November and December.

Gojjam's agroecology also comprised subecologies of low-lying wetlands that allowed substantial dry-season livestock forage. These black "cotton soil" bottomlands flooded during the summer months, preventing their use as cropland. Those wetlands, however, also offered habitat for mosquitoes, predominantly *Anopheles pharoensis* and *Anopheles arabiensis,* only the latter of which is an effective vector for malaria. Each season these areas were a sanctuary for cattle as well as a mosquito-breeding habitat. Most farmers avoided the areas at night, and nearby populations seem to have developed a limited immunity to the mild vivax malaria that appeared seasonally. Vivax malaria produced a low-grade recurrent fever in the few who chose to live near the wetlands, which most homestead sites avoided.[11] Malaria vivax, or "shivering fever," was a dreaded disease but not usually fatal. The much deadlier falciparum malaria was largely unknown in the region until recently. Burie land-use patterns over time created a biologically diverse setting that offered a balance of human settlement, plant communities, and an ecology of disease in reasonable control. Rapid changes in that balance, however, increased the risk of a disease outbreak.[12]

Residents of Burie had over time developed a folk taxonomy for their limited interaction with malaria, its causes, and its habits. Most of the Burie district extends over a rolling highland plain above 2,000 meters in elevation. Nighttime temperatures at that elevation were low enough to discourage mosquito breeding and parasite transmission. North and east of the main Addis Ababa–Gondar road, however, the topography becomes more elevated, up to 2,300 meters, with more pronounced upland vales favoring crops like barley, wheat, and potatoes. Residents confirmed the absence of mosquitoes in most of the district, certainly in the areas above 2,000 meters. As late as the mid-1970s neither mosquitoes nor malaria affected the midaltitude zones in Burie, probably because of temperatures too low for mosquito breeding, the absence of a sufficient human parasite reservoir, and inadequate vector density.[13] Burie residents' folk taxonomy of malaria's symptoms, cause, provenance, and prevention was a product of their limited exposure to it. Many of the generation of Burie residents over fifty recalled the disease as exclusively a scourge of lowlands and marshes. *Nidad,* as they called malaria, was a disease of intermittent high fever that afflicted people who passed through the Abbay Valley, especially Burie's Muslim merchants, or *jabarti,* who regularly traversed the lowlands between the major market centers in the empire.[14] Rarely fatal, *nidad* affected travelers after their return home, but the absence of the mosquito vector in the higher zones most people occupied meant that little if any transmission took place in the highlands themselves. Another malaria form recognized locally was the onomatopoeic *enqetqet,* or shivering fever, that sometimes affected those who slept outdoors, alongside their herds, at the edge of low-lying wetlands. But the most provocative association with malaria was a ubiquitous folk adage that linked malarial symptoms to maize. One informant recalled, with some amusement, "Our fathers used to tell us that to frighten us. 'Kids! Do not cut or ruin the tassels or eat the stalk of the corn or you will get shivers [enqetqet].' Our fathers said this to make us afraid. They said this to frighten us. The shivering fever."[15] Yemane reports that a fifth of his informants in the southern Zway region, more than six hundred kilometers to the south, expressed a similar belief of a link between maize and malaria (or at least recalled the adage), even though that area is linguistically and culturally quite distinct from Burie.[16] Most likely, this popular expression reflected the need to protect the valuable and tempting green ears from mischievous children, though people may have associated the ripening of the maize with the appearance of malaria in late August and early September, when the force of the rains lessened and nighttime temperatures rose above the crucial threshold of 15° to 18° C (59° to 64° F). It is important that the cultural memory of Burie residents (and those in Zway as well) made the association between malaria season and the more advanced stages of the maize plant's life cycle.

The Changing Agroecology of Maize in Burie

Until the 1980s, maize was, as mentioned earlier, a minor field crop in Gojjam, which appeared in the old African pattern—that is, mostly as a garden vegetable consumed green in the "hungry season" (August–September). For this purpose farmers chose from an array of early-maturing local maize types—the best-known of which were Mareysa, Harer, and Kafa—names that suggest a southern origin. Some of these probably arrived in Ethiopia via the Nile Valley in the nineteenth century, some originated with U.S. agriculture aid programs of the early 1950s, while others may have been descended from the original imports from the New World (via India) brought by Arab and Banyan merchants plying the Red Sea in the sixteenth century. Planted with the first rains, these early maturing types tasseled in late June or early July; as a vegetable, maize in Burie gardens was mature enough to consume by August or September. By the time malaria season came to the lowlands, then, the maize on the Burie plateau above was already ripening and the fallen pollen had been washed away by heavy July rains.

In the early 1980s, however, political changes initiated a transformation in Ethiopia's (and Burie's) agrarian economy and agroecological balance. Under the socialist military government known as the Derg, grain-marketing policy, forced labor, insecurity over land holdings, and government food security programs caused political disaffection but also changes in Ethiopia's national cropping patterns. Ethiopia's socialist government saw maize as a high-yielding field crop to replace labor-intensive teff and poor-yielding sorghum. As in the Soviet Union, the socialist planners saw maize as the ultimate product of an industrialized, scientific agriculture. For their part, farmers saw maize as a low-labor, quicker-maturing crop that provided food in insecure times when the socialist state sought to direct their labor to public works projects or to fix the prices for their other farm produce by means of the state marketing control boards. By the mid-1980s (unbeknownst to most) maize had unceremoniously surpassed teff and barley as the major crop produced in Ethiopia. Maize directly superseded sorghum on low- to midaltitude fields, replaced coffee in some areas of the south, complemented fields of chat in the southeast, and became the primary focus of state farms in the southern and western parts of the country.[17] In the Burie area itself two state farms in adjacent areas and four new producer cooperatives in an adjacent district introduced monoculture maize production, since that crop lent itself to large farm size, improved inputs, and mechanization of cultivation and processing. The programs that evolved during the 1980s used primarily such open-pollinated maize varieties as Alemaya Composite and A-511, which farmers could replant from their own fields rather than higher planting hybrids that were long-maturing and required purchases of new seed every year. By the mid-1980s improved open-pollinated maize had entered Ethiopia's northern agrarian economy as a major field crop.

In 1995 (after the fall of the Derg in 1991), the Ministry of Agriculture and the nongovernmental organization Sasakawa Global 2000 continued their infatuation with improved types of maize and began a demonstration package program to expand the use of inorganic fertilizers (urea and DAP), improved maize seeds, and agronomic techniques such as early row planting and intense early weeding in areas whose potential to improve national food production seemed high.[18] By 1998 maize had achieved 32.6 percent of the country's cereals production, more than any other grain. Between 1993 and 1998, the area of maize cultivation increased by 79 percent (from 808,900 to 1.45 million hectares—Appendix Table 8.1). In the northwest Amhara region, where Burie is located, the adoption rate was higher even than the national rate.

Nationally, the primary adoption emphasis with regard to maize variety had been on composites (or open-pollinated varieties), with only a few "hard-core" areas aggressively adopting hybrids. West Gojjam, however, was one of the primary zones to adopt a hybrid variety, Bako Hybrid 660, or BH660. The agronomic personality of this variety—long maturity, high yield, high stover (stalk) yield, and late tasseling—made it extremely popular with farmers but also made it central to the historical conjuncture and may well have resulted in the 1998 malaria epidemic.

The expansion of maize within Ethiopia's farming systems over the final decade and a half of the twentieth century has in many areas constituted an agroecological sea change in terms of labor calendars, links to national markets, and requirements for agricultural inputs—that is new seed, fertilizer, and storage. High yields and a new national market link significant wealth and increasing incentives for farmers to plant more hectares in maize, on as much land as possible. The result was a bold landscape transformation in the countryside and created new cropland for maize that came right to the doorstep of farm dwellings.

Terms of Change in Agroecology in Burie

While the new maize package program proved controversial in some areas of Ethiopia, farmers in Gojjam accepted the program readily because it offered preferred access to credit, fitted their new market orientation, and seemed to deliver yields substantially higher than those of traditional crops.[19] The significant changes of the 1980s at the national level in grain production trends, farm policy, and local politics had particularly deep effects in the Burie district. They signaled in many ways a sea change in agrarian life and geography, and in the agroecology of disease. The first feature of this change was a striking expansion of scale, ushered in by fresh road networks linking the district to national grain markets and sources of agricultural inputs (especially fertilizer and improved seeds). A major element in that road expansion was the completion of a road and a bridge linking Burie directly to the Abbay Valley and a new avenue for north-south commerce. A second element was the concentrated effort by the socialist military government in the 1980s to introduce agricultural development in the form of producer cooperatives, large-scale state farms, and advocacy of fertilizers, new crop varieties (especially maize), and village settlement schemes. Several of the efforts that met with massive opposition nationally actually achieved some acceptance in the Gojjam region. The military government collapsed in 1991, but after a few years' hiatus the new government began its own efforts at an Ethiopian Green Revolution through what they termed the "minimum package program," which encouraged farmers to adopt new seeds (predominantly maize), a regime of chemical fertilizer to create homogeneity in soil fertility, and modern techniques to increase yield on small farms.

Farmers in the Burie district in the post-1995 period, responding to this new opportunity with alacrity, dramatically increased their use of credit and nontraditional inputs. Burie's transition from an agricultural system based on mixed cereal crops to one dominated by hybrid maize had begun in the postrevolutionary period of the 1980s with the establishment of two model state farms, a tractor mechanization facility, and several producer cooperatives for raising improved maize with the aid of both mechanization and fertilizer. With the collapse of the Derg's socialist programs in the late 1980s and the ascension of the post-Derg government in 1991, a new phase emerged in Burie, as it had nationally. Formally launched in 1995, this national program took a strong hold in Burie, transforming the district from one renowned for its teff and finger millet to one increasingly dominated by the cultivation of improved maize grown as a field crop, especially the late-maturing hybrid BH660 (Appendix Table 8.2). This maize variety produced prodigious yields of up to six to eight tons per hectare in conjunction with Gojjam's red clay soils and consistent months of rains in the cropping season that stretched from sowing in late April to May until the harvest in November or December.

The final element in this agroecological change was the global and regional climate fluctuation that had produced drought in many areas of Ethiopia but that had provided the Gojjam area with several years of abundant and consistent rains, raising on-farm production and encouraging farmers to accept offers of credit for agricultural seeds and fertilizer. Even absent dramatic increases in temperature, the somewhat heavier rains in 1998 did not demonstrably lower ambient nighttime lows or daytime highs. In this regard, 1998 was a year of sustained rainfall that extended well into the fall months. Overall, neither temperature nor rainfall would seem in and of itself to account for the precipitate shift in malaria cases, given that climate records reveal no significant anomalies in either rainfall or temperature. Paradoxically, however, the consistency in climate over the 1996–1998 period no doubt encouraged farmers to choose the high-yielding, but long-maturing BH660. Few Burie farmers had had recent experience with rainfall deficits.

The proliferation of maize in the Burie area over the nine years following the 1995 introduction of the hybrid-maize package was remarkable: the area planted to maize more than doubled, raising the percentage of the total area devoted to maize from 21 to 36 percent. While the figures on maize's dominant role in cereal production reflect a national trend and document its local manifestation in Burie, they mask the even greater dominance within Burie. The final figure of 26,758 hectares planted in maize for the 2001 crop year indicates that in Burie itself maize accounted for well over a third of all cereals and was growing rapidly overall. For the 1998 crop year, it was certainly at least that high. Moreover, for the purposes of this research the key issue is the high concentration of maize in eight of Burie's twenty-one localities—where maize's share probably reached 85 percent of the all cereal planted.[20] One farmer commented to me that "if a thief hides in a maize field beginning in Gulim he can run under cover of that maize until he comes out at Fetam Sentom"—that is, twenty kilometers to the south![21]

Burie farmers' heavy concentration in maize cultivation brought with it a significant change in the area's pattern of human settlement, indicative of the emerging modern agroecology of maize. In those localities most intensely invested in maize, farmers had altered their patterns of labor, fuel supply, livestock forage, and domestic space. The formerly dispersed homestead sites on which fenced garden space separated houses from field crops have given way to homesteads gathered in small nucleated villages or towns of rectangular houses with corrugated iron roofs. Those settlements include grain storage buildings, teahouses, and shops: all signs of the increasing income base of rural inhabitants. Most remarkably, the houses in these villages have no defined *gwaro* with maize cultivation running up against the walls of the houses themselves. Some plantings stood within half a meter of the house, and virtually all the dwellings we surveyed had maize fields within ten meters of the dwelling—indeed, most were much closer (that is to say, they fell well within Yemane's criterion). When asked about the disappearance of the fenced house garden, a farmer laughed and, pointing to a pile of straw adjacent to the house, said, "You see that stack of teff straw there? Once the cattle finish that one we are going to plow up that land too"—for maize.[22]

The physical appearance of these new villages in maize locales gives a spatial expression to a more profound shift in material culture. Farmers in the area insisted with enthusiasm that maize touched all aspects of their lives, for both genders. Maize stalks provide cooking fuel, green leaves supply fodder for livestock, young ears offer snack food, women now mix maize flour into injera (Ethiopia's distinctive teff pancake) and wheat bread batter. Shelled cobs are put to myriad uses around the household. Instead of the formerly ubiquitous stacks of teff straw for dry-season fodder, virtually all houses in these areas sport a lean-to stacked neatly with dried maize stalks preserved for fuel. Teff straw is now less a farm by-product than a commodity purchased at the market with the profits from bulk maize sales, an interesting measure of agricultural development.

The end result of this conversion to maize approaches what food historian Betty Fussell in the historical context of the American Midwest and Italy's Veneto calls "corn madness."[23] Yet, these areas of high maize concentration still stand in marked contrast to the low-maize Burie district where other cereals predominate because of soil type, elevation, or remoteness from a road. In those places, traditional fenced household gardens are still common, as is the cultivation near the homestead of potatoes, kale, and squash.

A chief catalyst in the agroecological change appears to have been the enormously productive hybrid-maize variety BH660, locally called Silsa Sidist. Released in 1993 and introduced to the Burie area beginning in 1995, by 1998 BH660 had replaced virtually all other varieties in the area.[24] In the Amhara region as a whole, the percentage of farmers adopting improved maize increased from 16 percent in 1995 to 43 percent in 1998. A 1998 survey of the entire Amhara region indicated that 80 percent of the farmers sampled had adopted improved maize, with BH660 the most popular variety overall, and virtually the exclusive choice in Burie. In that same 1998 survey, farmers identified their top four reasons for choosing improved maize:

High yield
Tolerance of lodging (stalk collapse)
Better germination
Quality and higher yield of leaves and stalks for fodder[25]

The maize type and its specific characteristics contributed directly to the agroecological change that underlay the 1998 malaria epidemic. Most important, BH660's tasseling and pollen release take place later in the season than in older varieties—that is, in late August or September, at precisely the time when temperature and moisture are ideal for mosquito breeding—a confluence in the agroecological calendar that evidence suggests has serious implications for malaria infection.

The 1998 Malaria Epidemic in Burie

The number of malaria cases and deaths in 1998 was unprecedented in the region. Previous disastrous malaria epidemics in Ethiopia (in 1953 and in 1958) had affected wide areas of the country and the Lake Tana region, but had not reached Burie. In 1998, by contrast, in the Amhara region alone 3.4 million people were affected and more than 7,700 deaths occurred.[26] The outbreak in the Burie area began in May and June in a few districts adjacent to those where malaria is endemic but then expanded into higher altitudes in September–October 1998, only to gain additional ground in October and November when the powerful wave of malaria washed into previously malaria-free zones. The outbreak subsided in January 1999, leaving victims and officials to wonder about its causes and consequences. That the dramatic August–September 1998 spike coincided with the BH660 tasseling time suggests a correlation between those two events, and it is borne out by the laboratory evidence.

Yemane's 2000 research findings in a controlled experiment on the connection between maize and *Anopheles arabiensis* beg the question of whether actual ecologies on the farm, in rural settings, and in maize interaction produced increased *vector density* and *rates of inoculation,* thus providing key components of the conjuncture that underlay the virulent malaria outbreak. The 1998 malaria epidemic in Burie offers an ideal case for investigating the role of maize in malaria transmission. Evidence of the timing of pollen release, the presence of increased breeding sites for larvae, and the proximity of sources of pollen to human housing strongly suggests that the recent agroecology of maize and malaria in the Burie district has supported an increase in vector density, the range for mosquito habitat, and the bite rate, all factors leading to an outbreak of epidemic proportions. The larger size and greater longevity of maize pollen–fed *Anopheles arabiensis* suggest that the presence of pollen allows a greater number of malaria parasites to complete their extrinsic life cycle and therefore gives rise to a higher rate of inoculation and transmission (that is, more and longer-lived mosquitoes carrying deadlier parasites).

There are two levels of evidence to consider. The first is the statistical association between the presence of maize and its chronological and spatial coincidence with the disease in those summer months of 1998. Most remarkable are the case data from local clinics and health centers from August to December 1998 and an assessment by the Ministry of Agriculture's Burie district office that ranked agricultural locations according to whether maize cultivation was of high, medium, or low density.[27] When the factor of altitude is held constant (since that affects temperature), the impact of heavy maize cultivation on malaria infection is shocking. In areas of high maize production and proximity of pollen sources to mos-

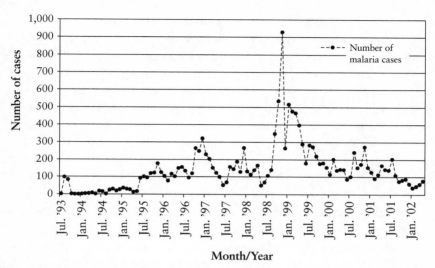

32. Monthly malaria cases in Burie, July 1993 to January 2002.

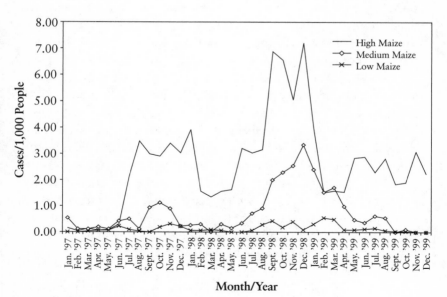

33. Malaria cases in Burie.

quito-breeding sites, 24 percent of the population registered positive diagnoses for malaria (falciparum and vivax). In areas of medium maize production, 9.2 percent came down with the disease. And in areas of low maize production only 2.5 percent of the population contracted malaria. The ratio of cases per thousand population in areas of high maize growth to those of low maize growth was 9.5 to 1.0. In other words, farmers who grew maize in Burie's midaltitude zones were approximately ten times more likely to contract malaria than were those who cultivated other crops in the *same* altitude zone. The correlations by themselves do not necessarily imply a cause-effect relationship, but given the strong scientific links proved by Yemane's laboratory work, the weight of this evidence is significant indeed.

More particularly, the conditions in the Burie district in summer 1998 fulfill virtually all three of the criteria that Yemane's entomological work hypothesized. First, areas of high maize production shared the common (and unprecedented) settlement pattern in which houses are within ten meters (and much closer in the majority of cases) of tasseling maize. Zones where little maize was grown that had household gardens used to maintain considerable space between the dwellings and field crops. In those areas maize is no longer as ubiquitous as a household garden crop, though it appears here and there. The late-tasseling characteristic of BH660 maize brought pollen-producing fields into close proximity with mosquito-breeding habitats at precisely the ideal time (early September), when the torrential rains begin to subside and stagnant pools offer attractive sites for egg-laying. Moreover, human dwelling places offer temperatures a few degrees higher than the ambient ones and as such are attractive to mosquitoes. Such conditions result in a high density of adult mosquitoes in close contact with humans. These were the conditions that led directly to the malaria epidemic of 1998.

Second, the higher survival rate of larvae to pupae and therefore adults conforms to oral testimony that mosquito populations in 1998 were especially dense and appeared for the first time inside farmers' homes. Third, farmers anecdotally reported a particularly high rate of biting in homes and around fields. A higher bite rate would mean a higher likelihood of malaria transmission. Finally, the increased mobility of populations, which were moving from zones where malaria was endemic to zones that were previously malaria-free, was a product of improved transport, trips to market, and the growth of human populations in new village settings densely planted with maize. This last point indicates the presence of a human-parasite reservoir that increased in size as the epidemic progressed.

The agroecology of maize in this case is related specifically to the agronomic characteristics of BH660, in that its high yield has encouraged an expansion of its planting within close proximity of dwellings and breeding habitat. Moreover, the tall stature of the hybrid plant (255–290 centimeters) and the extension of tasseling time into early September places the maize in much closer contact than ever before with the warming temperatures that also encourage mosquito breeding. Conditions on farms in Burie approximated the ideal conditions for mosquito propagation indicated in Yemane's controlled study. Thus a correlation between the new agroecology of maize, human settlement, and vector density seems justified in explaining both the intensity and the geographic expansion of the malaria epidemic. Epidemiological evidence also strongly indicates a correlation between the origin and progress of the 1998 epidemic and the specific areas of intensive maize cultivation. This circumstantial evidence relates the intensity of cases reported in Ministry of Health centers and clinics to those areas with the highest percentages of land planted in maize.

Secondary factors of economic change have also contributed to genesis of the 1998 epidemic in Burie. These include the 1988 opening of the Burie-Nekempt road that brought a human population that served as a parasite reservoir into contact with highland zones, the national agricultural policy of promoting improved maize, and the overall effects of global climate change in the form of rising highland temperatures above the threshold for the development of the falciparum parasite (17–19° C, or 62–68° F).

The unintended consequences of economic development and environmental change often include new disease effects on humans, plants, and livestock. The link between schistosomiasis and irrigation, for example, is already well established. Each of the features of the 1998 malaria epidemic in Burie contributed something to the historical conjuncture that resulted in high mortality figures and expanded the geography of infection. How often such conditions will replicate themselves is an open question. Some of the evidence for the link between maize cultivation and malaria is circumstantial, but the additional testimony of lab results is compelling. Taken in concert with Yemane's controlled study of maize pollen–vector interaction, the results must be considered very seriously. The evidence presented here does not suggest that maize cultivation is a proximate cause of the 1998 malaria epidemic, but rather that it is part of a combination of events of climate, agroecology, and human settlement that resulted in the outbreak of the disease, its geography, and its intensity. If the 1998 epidemic began in low-lying areas where malaria was endemic, maize cultivation was the fuel that propelled the disease onto unfamiliar ground in the highlands.

Obviously, the cultivation of improved maize has an important place in increasing food supply in such chronically food-deficient areas as Ethiopia. But this study provides evidence that the increasing dominance of maize, along with other factors, such as population mobility and an increase in global temperatures, makes epidemics more likely in the future, whenever conditions and timing are favorable.

The evidence on maize-malaria interaction strongly suggests that public health programs must make prophylactic measures and habitat control an integral part of extension programs to increase maize production. Meanwhile, national and international agricultural policy suggests that at least one important element of epidemic malaria—maize—is expanding globally in popularity and range, and nowhere more than in Africa.

Maize as Metonym
in Africa's New Millennium

9

In the 1999–2000 edition of the publication *Maize Facts and Trends,* economists Prabhu Pingali and Shivaji Pandey used economic projections from the International Food Policy Research Institute (IFPRI) to argue that a radical change was under way in the world demand for cereals. By the year 2020, they argue, world demand for maize will surpass the need for both rice and wheat. The global demand for maize will increase over the first two decades of the twenty-first century by 50 percent, from its 1995 level of 558 million tons to 837 million tons. For Africa, where humans eat more than three-quarters of the maize produced (and if South Africa is excluded, more than nine-tenths), the annual demand for maize will virtually double, to 52 million tons. Table 9.1 shows the global trend as projected by economists, and Africa's place within that.[1]

From an economist's perspective, such projections are necessarily simplistic (assuming finite variables) and possibly based on narrowly conceived notions that ignore the complexity of past events. It nevertheless seems true that maize will be, for better or worse, center stage in determining future food supply, farming systems, and social production—globally, but especially in Africa. This prominence suggests to optimists that maize will be an engine of African economic growth, though the evidence of the past also makes it plausible that it may prove an unsustainable folly. Maize

Table 9.1 Global maize demand projections, 1995–2020

Region	1995 demand (in millions of tons)	2020 demand (in millions of tons)	% change
Global	558	837	50
Developing world	282	504	79
E. and S.E. Asia	150	280	46
S. Asia	12	23	92
Sub-Saharan Africa	27	52	93
Latin America	76	123	62
W. Asia, N. Africa	16	26	63

Source: P. L. Pingali and S. Pandey, "Meeting World Maize Needs: Technological Opportunities and Priorities for the Public Sector," part 1 in P. L. Pingali, ed., *CIMMYT 1999–2000 World Maize Facts and Trends. Meeting World Maize Needs: Technological Opportunities and Priorities for the Public Sector* (Mexico City: CIMMYT, 2001), 1.

has influenced virtually all areas and eras in Africa over the past half millennium, for it is in breadth of geographic distribution and total production Africa's (and the world's) most widely adapted food crop, being cultivated in a dizzying range of altitudes, moisture conditions, soil types, ecologies, and farming systems. By contrast, another New World crop, cassava, also has a wide range and a high yield per hectare, but it has nothing like maize's ability to appear as a major field crop in one setting and as a garden vegetable in another and to dominate one's visual impression of an agricultural landscape. The presence of maize in many ways defines Africa visually, nutritionally, and economically.

The sweep of maize's life course in Africa over the past five hundred years is emblematic of the continent's economic, social, and ecological history—for instance, in the ways both histories touch on deepening engagement with the world economy, tardy but rapid urban growth, and the willingness to adopt novel ideas and forms of material life. Africa's cultural and physical landscapes in the late twentieth century were the cumulative expression of these historical forces.

What are the central themes that characterize Africa's encounter with maize? Two come to mind. The first is African men and women's appropriation of the crop across the wide range of cultural, economic, and ecological settings. Particularly important to this process is that African agriculture in most of its forms was an artisanal rather than an industrializing activity. The vagaries of climate, old soils, and tropical ecology meant that African farmers (men and women) juggled crop mixes over time, on different plots, and through the changing seasons. Their goal was to choose the crop whose characteristics best suited the ecology of the field they intended to cultivate. In most cases African maize fitted into an existing cropping system, often filling an unoccupied niche for an early-maturing crop with minimal processing requirements. In post-Columbian North America and Europe agriculture moved on the whole toward an industrial model whose logical endpoint was monocropping; farmers in this modern era have sought a homogenous crop to fit fields that, thanks to chemical fertilizer and herbicides and tractor cultivation, were predictably similar to every other one. In most of Africa that outcome has been an uncomfortable imposition only of the late twentieth century.

We can trace life history of maize in Africa through its gradual and culturally specific transformation from a vegetable into a staple cereal. This change in use and agronomy took place in different times and settings across Africa. In semihumid zones in West Africa or southern Africa, maize offered clear advantages in yield and labor requirements over the indigenous African cereal crops of sorghum and millet. When rainfall exceeded five hundred millimeters per year, maize replaced these indigenous crops, though it supplanted rice less completely, in places like the Upper Guinea coast and around Akyim in Ghana.

Interestingly, it also replaced sorghum—and more significantly wheat—in northeast Italy. But in the Italian case, peasant farms moved directly to small-scale monoculture of maize cereal, with a consequent and infamous descent into the ravages of pellagra, caused by chronic vitamin deficiency. Within two generations, however, Italian cultivators and agriculture had graduated to a capitalist industrial scale tied directly to world markets. In the West African case, maize in most forest and semihumid ecologies became the dominant grain within a century of its introduction, but, as we learn from historical testimony and modern practice, it flourished in forest areas, where it assumed an important role within a tight-knit crop mix of tubers, vegetables, and minor cereals. In highland Ethiopia, by contrast, a wider range of annual cereal crops was available; there, maize's niche was a narrow one that for several centuries relegated it to a garden snack, a minor player in the diverse mix of native African cereals and pulses. Over the long term, however, maize has settled into its status as a cereal staple.

In these historical agroecologies, farmers each season selected from their harvest the maize offering their favored colors, flour textures, and field characteristics (height, husk cover, early maturity, disease resistance), thus shaping the next generation's crop to

34. "The National Stomach," billboard, Limpopo Province, South Africa, 2004.

their specific needs and conditions. By culling through and saving seed, farmers constructed specific genetic qualities in different regions, villages, and cultural geographies. During the first centuries of maize's peregrinations within Africa, farmers were able to choose from flints, dents, floury types and a kaleidoscope of genetic variations within those types and assert their own aesthetic and agronomic sensibilities to shape the results. These genetic personalities (genotypes), of course, built upon what New World farmers' selections had engineered into those seeds generations before they found their way to Africa. The observers who first came across and recorded the progress of maize in its African surroundings found a stupendous set of genetic building blocks: floury maizes in Gold Coast forest zones, hardy flints in dry areas cultivated by Nguni in the Natal sweetveld, early-yielding flints in Ethiopian homestead gardens, and high-yielding American dents on the commercial farms of the Cape, Transvaal, and Orange Free State. The list goes on.

Moreover, the social implications of maize were also quite varied. It seems likely that women nurtured it as a garden vegetable crop in Ethiopia; once it had become a field cereal crop, it entered men's control. In Northern Rhodesia's Tonga region, by contrast, maize's marketing potential as a field crop in the 1920s and '30s placed it rather firmly in the male domain. In Western Kenya, Luo women were the innovators in exchanging maize seeds and managing the crop's place in the overall livelihood strategy for the household.[2]

Politics were never far from the maize field. Within settler colonies of the twentieth century, such as Kenya and Southern Rhodesia, race was a defining criterion. Control over maize as a cash crop became a central concern of the colonial state almost as soon as colonial economies took shape in the interwar period. Colonial governments in the Rhodesias and Kenya established maize marketing boards to support the prices of maize grown by white settlers, and restrictive legislation to block or at least restrict market access for black African maize farmers. In the view of the colonial settler state, native Africans were needed as miners or agricultural laborers, and the settlers and state policy makers often viewed Africans' growing of maize for the national market or even for subsistence as a subversive activity that threatened the underpinnings of the settler economy. At the same time, outside the settler context—for example, in Malawi—African farmers (in this case women) defined and defended their own preferences for local types of maize against the postindependence state's attempts to dictate to farmers or cajole them into accepting new types whose storage life, flour texture, and taste did not appeal to them. In most places and times African farms retained within reach an intimately nurtured variety of local maize types as a means of managing risk from climate or disease or simply of expressing their individual and collective aesthetics of taste and texture.

This early historical phase of the Africanization of maize expressed itself by place, ecosystem, economic culture, and aesthetic preference in types as heterogeneous as the regions of the continent itself. The "craft" of African agriculture as practiced across Africa's many landscapes made this inevitable. While it is possible to generalize about particular agroecologies (the Ethiopian highlands, for instance) where a broad consensus emerged on how and what to cultivate, in other cases (such as Yorubaland) color and texture preferences, village-level nuances of style, and microecologies promoted greater diversity.

The second discernible phase of maize's African journey and acculturation was characterized by an inexorable impetus toward standardization of cultivation methods and of the genetic expressions of the maize plant itself. This homogenization was largely a product of the twentieth century and the progression of Africa's political ecology from predominantly local initiatives to colonialism to globalization—a process in which Africans have nevertheless continued to occupy a distinctive place in the world of maize cultivation, albeit on new terms.

As a cultural expression and in regional cuisine, African maize is like no other. The recent standardization of maize cultivation in Africa has set Africa apart from other world areas. Maize plays a critical role in African diet. As has often been stressed in earlier chapters, African maize serves overwhelmingly as human food (not fodder, as is the case in Europe, North America, and Asia). Only in South Africa and Tunisia is maize primarily used as livestock feed. It could be said that Africans' preference for white maize, as opposed to the yellow types favored by most of the world, is a measure of the integration into an urban wage economy, as Africans have become workers, bureaucrats, and market-oriented farmers. Local variegated maize is now not merely a curiosity, but also a marker of poverty outside the full grasp of globalization. The old maize types command little research or support from the state or international agencies. Even modern seed banks that otherwise recognize local knowledge collect the multicolored maize only haphazardly.[3] Local maize grown as a subsistence crop rather than in its industrial hybrid form is also distinctively though not exclusively African (Central America and some of Asia still exhibit this phenomenon).

Maize as an industrially produced cash crop appeals to the global system because it is controllable by the state and corporate agriculture and amenable to economies of scale in cultivation, processing, and in research investment. It thus suits global economic forces that seek increased food production, the circulation of commoditized agricultural inputs (fertilizer, herbicides, and pesticides, genetic modification), and a product that will be comparable across geography and cultures. But it also appeals strongly to forces of political control and centralization. From socialist Ethiopia and Stalin's Soviet Union to apartheid South Africa and the Sasakawa Global 2000 Agricultural Project, state planners and agricultural empiricists alike have recognized the appeal of maize. James Scott has described "high modernism" or "legibility," which is to say the quest for predictability and economies of scale; maize

35. King Korn, KwaZulu Natal.

is the ultimate "legible" food and crop, one that holds attraction for ambitious governments enamored of large-scale projects.[4]

Can maize also be a force for democratization? Perhaps. The prevailing homogenization of maize over the course of the late twentieth century has coexisted with a dissonant subtheme: the resistance of local preference. It is a reminder of the rigors of subsistence in areas where market forces are weak and farmers must hedge their bets against a cruel climate and economic isolation. In some places maize as a subsistence crop expresses the intransigence of a structural poverty, and in other places its presence may be a sign of a more genuine recalcitrance of local economic culture and taste, an attempt to hold back the tide of global change led by multilateralism, global culture, and the efficiencies of scale.

Global Maize as Jurassic Park?

Global trends in the cultivation and processing of maize as human nutriment, cattle feed, and industrial raw material emphasize yield and plant characteristics that are responsive to chemical fertilizers, herbicides, and pesticides.

Monoculture's attraction lies in efficiencies of scale that fit macroeconomic models and corporate spreadsheets but risk the kind of biological chaos and unintended consequences that pop-science novelist Michael Crichton attempted to illustrate in his *Jurassic Park*. In that imaginary scenario, the biological impetus toward procreation and chromosomal leaping causes Hollywood dinosaurs to run amok. In crop production the unpredictable and unintended consequences of monoculture loom on the African horizon and may affect the apparently boundless potential of maize. The description in Chapter 8 of the accelerating effect of maize on the recrudescence of malaria and its expansion into the Ethiopian highlands presents at least one real-world consequence of monocropping hybrid maize in African settings.

Yet the role of such private agribusiness seed companies as Archer Daniels Midland, Monsanto, and Pioneer in the research and development of new maize types is critical. The most important issue for the future is whether those corporate planners perceive that they can address the needs of Africa's farmers and still remain profitable. Doing so would probably require adopting a long-term perspective and a new definition of profitability.

Another concern is the increasing risk that derives from mycotoxins (specifically aflatoxin). Mycotoxins are chemical substances naturally produced by certain species of fungi that grow on crops in the field and in storage. Maize is one of those crops affected by mycotoxins in general and by the fungus *Aspergillus flavus* in particular. *Aspergillus flavus* produces the poison aflatoxin, which damages DNA in humans who ingest it and is the strongest known chemical liver carcinogen. Aflatoxin is also synergistic with Hepatitis B and C (common afflictions in Africa), and in combination they raise the risk of liver cancer to ten times the rate for these carcinogens taken individually. Recent work in West Africa suggests the further possibility that aflatoxin suppresses the human immune system, in an effect similar to that of HIV / AIDs. Yet another mycotoxin, fumonisin, has appeared with maize in southern Africa and is being studied now on the eastern Cape for its role in esophageal cancer. Aflatoxin produced in stored maize also affects livestock and thus the human food supply.[5]

The appearance of aflatoxin most often results from the improper storage of maize or groundnuts. In most developed countries the presence of the mycotoxin is monitored commercially and presents little danger to consumers. A genuine risk exists in developing countries, however, where on-farm and commercial storage are unregulated. Recently, maize consumers in Thailand suffered from the often deadly effects of aflatoxin due to poor maize storage. The danger to African maize consumers is almost certainly greater, given the high percentage of African maize consumed by humans. Recent research indicates that in conditions of drought stress, mycotoxin-producing fungi from the soil invade plant roots and increase the presence of the toxin in the grain that reaches storage.[6] The frequency of African droughts in the past three decades clearly indicates that Africa is more vulnerable in this regard than any other area of the world. In Central America the aflatoxin risk is less, simply because historically it has been common practice there to treat maize kernels with alkali (lye) in the course of food preparation.

In 2001 a joint committee of the World Health Organization and the Food and Agriculture Organization stated that for some or all of their lives Africans would be exposed to dangerous levels of aflatoxins. The committee also found that in Africa average exposure to the mycotoxin fumonisin exceeds the tolerable daily intake and that for about 10 percent of the population, its levels exceed those limits by 300 percent.[7] Small-scale farmers—that is, the vast majority of Africa's maize producers—are subject to the gravest risk, since they most often consume grain they stored themselves, having marketed the better-quality maize.

The unintended biological threats to African maize consumers increase in intensity under a system of maize monocropping. These threats have solutions, but addressing the problem requires research and policy decisions that would call for governmental reallocation of scarce health and agricultural resources. Considering the plethora of other health menaces in Africa, that is unlikely in the near future. We can thus anticipate some unexpected biological outcomes from the enormous commitment to maize production foreseen in the IFPRI projections of demand for maize mentioned at the outset of this chapter.

The potential for unintended consequences from intensified maize cultivation in Africa suggests how policy makers might approach the thorny issue of genetically modified maize. In the case of genetically modified *Bacillus thuringiensis* (Bt) maize, where the maize produces its own pesticide, global forces have expressed a clear preference for the legibility of monocropping, rather than an approach that produces many types for use in a wide variety of ecological situations. The message from international agribusiness and African ministries of agriculture is that they have a fundamental commitment to monocropping—a choice strongly at odds with Africa's historical cultivation patterns.

There is little if any direct evidence that GM maize threatens human health or damages local ecologies. Yet evidence is also insufficient to enable scientists to conclude that it does not.[8] This conundrum is a global issue, currently resolved on the basis of ideology rather than evidence (with Greens lining up against crop geneticists and profit-seeking corporate managers). Thus politics and ideology, rather than science or African farmers' choice, will probably determine whether genetically modified crops are to be a feature of Africa's agricultural landscapes in the twenty-first century. Research is unlikely to resolve the issue in the near future, but the widespread adoption of GM maize or other new varieties as an answer to Africa's food needs would doubtless entail unintended consequences. Public health institutions and agricultural science will need to anticipate these outcomes, or at least to address them as they present themselves.

Another, more tangible factor that will shape the future of African maize is the state of both the global and local economies as they affect world trade in maize. While Africa is strongly affected by the global trade in maize, it has little impact on it. The United States alone exports more maize in an average year than Africa's total production. International prices for maize are determined by exports from developed countries (such as Argentina and Brazil). China's admission to the World Trade Organization (WTO) will have a significantly larger effect on world maize markets than will any action taken in or about Africa, since it will add to the WTO roster a maize-producing economy far larger than Africa's.[9]

In the period from 1996 to 1998, for example, Africa as a whole (including North Africa) was a net importer of maize, its most important food crop. Eastern Africa and southern Africa were overall very minor net exporters of maize (127 tons), but those exports were exclusively attributable to South African and Zimbabwean commercial farms. Excluding those two countries, southern and eastern Africa imported 1,162,000 tons annually in 1996–1998. West and central Africa imported 184,000 tons, and North Africa imported a whopping 4.89 million tons (Egypt accounting for 2.85 million of that amount). A few African countries either export annually (in a year of good rainfall) or hope to do so in the future, as projected in their long-range economic planning. The Ethiopian government, for example, has expressed its intention to be a net maize exporter soon as a measure of the success of its maize extension program. With the exception of South Africa, however, the prospects for maize export are bleak, if only because Africa's costs of production and transportation far exceed those of industrialized producers, such as the United States, Argentina, and Brazil. In Ethiopia alone, transportation costs exceed $175 per ton, making export uneconomical, despite the symbolic value food exports have for that government. Maize delivered to Kenya or Ghana from Minnesota or the Argentine pampas is considerably cheaper than is maize coming from Djibouti (Ethiopia's major seaport) or Maputo.[10] Moreover, the demand outside the continent itself for Africa's white maize is minuscule, and that small demand is met by South Africa.

Can maize therefore be Africa's economic engine, its saving grace? An increase in African countries' domestic food supply would no doubt be a blessing, but for most African national economies, increasing their maize production for export would be equivalent to a strategy of producing more dearly at home a commodity that it costs less to import—a national economic strategy abandoned in the 1980s. Unlike potential nonfood exports, however, maize is also a crop that increases food security at home, as long as domestic infrastructure is adequate to redistribute it.

A real cautionary alarm should be sounded for a continent that is likely, in the face of global climate change, to be subject to increasing drought, biological calamities, and the vagaries of the international markets for grain and agricultural inputs. It is a gloomy prospect.

Are Maize and Africa an Ideal Match?

There is an attractive and more optimistic counterargument to the above factors. Maize is ecologically well suited to Africa because of the wide range of agroecologies there, from highland plateaus to humid tropics and semiarid subtropics. Thanks to its genetic diversity and malleable nature, maize has accommodated itself to virtually all those conditions. Its overall success in Africa is confirmation of the general rule that plants of economic value thrive the more, the greater their distance from their place of origin, where disease organisms are as fully evolved as the crop plant itself.[11] More important, Africans have accepted maize into their diets; it is now an intrinsic part of national cuisines and tastes. Even recalcitrant Ethiopia has shown some evidence of increasing its maize consumption, albeit slowly, within its own alimentary and brewing traditions.

A less self-evident reason for Africa's suitability for maize is its largely untapped powerhouse of local maize genetics. Global agricultural science seems to have overlooked this advantage, though Norman Borlaug's argument about a fossil gene resisting the American rust is worth noting. It was not until recently, however, that the CIMMYT headquarters in Mexico learned that its gene bank contained no African materials; only New World cultivars were collected and preserved there. Although this policy changed recently, a systematic effort must be made to preserve this uniquely African heritage. Though maize is a New World crop, its genotypes have continued to evolve, as we have seen, through farmer selection, the serendipity of mutation, and the natural selection process brought about by disease and pests. Douglas Tanner, a CIMMYT breeder based in Addis Ababa, has even speculated that the high ultraviolet concentrations on the Ethiopian plateaus promote a high rate of mutation that serves as a source for new crop organisms. Might the same process have created unknown plant cultivars as well? It seems quite likely that African adoption of maize over the past half millennium has produced some novel genetic combinations and successful adaptation to local ecologies that will contribute to the future possibilities for maize elsewhere around the globe. Unfortunately, the weight of genetic homogenization in the past half century has probably reduced much of the maize biodiversity that Africa once enjoyed.

To date, investments in African maize have focused on high yield as the premier trait to develop. Other desirable characteristics, such as drought resistance, disease resistance, and now quality protein content, have emerged only on a minor scale. Breeding programs are valuable but have largely used non-African genetic admixtures rather than searching for African plant materials from the points of historical dispersal on the continent. Major efforts to improve smallholder maize production, such as Sasakawa Global 2000, have promoted "off-the-shelf," cookie-cutter varieties that require expensive and potentially hazardous chemical inputs, rather than promoting the search for existing genetic diversity within Africa and Latin America.[12]

The real challenge is for private seed companies to form partnerships with public organizations to address the needs of Africa's small farms and consumers. Innovative breeding programs are now under way that add genetic diversity to African maize that tolerates drought and low fertility. Some private seed companies and national programs are already responding to this trend as an economic opportunity.

Maize and the African City

Two major themes in twentieth-century African history are the delayed but now rapid growth of Africa's urban settlements, especially district-level towns and major urban centers, and the expan-

36.
African
urban
maize,
Accra.

37.
African
urban
maize,
Bahir
Dar,
Ethiopia,
2003.

sion of maize as the primary component of Africa's food supply. Are these two trends compatible or contradictory? In fact, the evidence suggests a confluence, or perhaps a synergy, of the two trajectories. Africa has had a rich history of urban settlement around trade centers along trade routes (Timbuktu, Kano, Ibadan, Benin City), coastal enclaves (Mombasa, Zanzibar, Massawa), or more recent cities that have their origins in colonial administrative centers (Nairobi, Harare, Dakar, Kinshasa). In comparative global terms Africa is distinctive in its patterns of urbanization. In 1900 it was the least urban continent (around 12 percent cities), but by the late twentieth century it had become the most rapidly urbanizing land mass.

Africa's urban centers are thus affected not only by surging

38.
Maize consumer item, Zimbabwe, 2003.

growth, but also by very recent global trends in migration, capital remittances, and agricultural trade. Because the expansion has come late on this continent in relation to other world areas and has coincided with economic globalization and the weakening of postcolonial states, African urban growth has its own distinctive patterns. Africa displays a number of innovative features, including the continued circulation of people, capital, and resources between agrarian and urban settings, rather than a unidirectional transfer. The impact of this circulation appears in what might be called the African urban metabolism. This metaphor of the body refers here to the circulation of natural resources, including water, waste, and food, within and through the city. The realities and dynamics of this process are only poorly understood, for they often take place outside formal planning structures.

Africa's urbanization has to a large degree paralleled its growing dependence on maize as a food source. In southern Africa the direct links between urban growth, mine labor, and maize are most obvious. Maize has also surfaced in Africa's new and vibrant urban settings, now that new forms of cultivation have appeared within dense human settlements. Urban agriculture may be, in fact, a new form of artisanal farming, since it often employs multicropping and combines elements of market gardening, household subsistence farming, and experimentation with intercropping. Urban agriculture, in fact, may offer a new set of possibilities for maize and intercropping, in that it adapts itself readily to market gardening. The high market value of intercropped vegetables and fruits means that they generate cash income as well as household food.

Maize as an agricultural raw material presents a prospect for value-added industry on a small to medium scale. In Ghana more than 80 percent of maize-producing households purchase processed maize for their own consumption because of the specialized skills involved in preparing food maize. Those food purchases help generate income locally, especially for women. In Ethiopia, where small district centers have been the fastest-growing urban settings, maize marketing has provided substantial private-sector employment around trading and service. Processing of maize into food and industrial commodities (starch, infant food supplements, oil) would provide a small industrial base and accelerate the process of converting maize into part of the urban food supply for the expanding middle class. Agriculture in Africa's urban "halos" has been the subject of social science research, but very little agricultural investigation has been done on urban crops or the agronomy peculiar to them.

Periurban agriculture is an essential part of urban growth in places like Kinshasa, Accra, Ibadan, and Addis Ababa, where poor rural infrastructure has created strong incentives for urban dwellers to engage in intensive market gardening within the city's environmental footprint. African urban population includes a burgeoning middle class, especially in the service sector, whose increasing demand for meat is integral to the expected demand for maize in the next two decades. The International Food Policy Research Institute assumes that the bulk of that increasing demand for poultry and livestock will take place in eastern and Southeast Asia, but the growth of livestock-fattening industries in places like the periphery of Addis Ababa and other major urban centers in recent years suggests that African cities will continue to combine wrenching poverty with increasing consumption.[13] Corn curls, long a junk food staple in South Africa and Lesotho, are now produced locally in Ethiopia.

The case of Rhodesia's SR-52 suggests that maize also serves a political purpose. Maize is doubtless going to continue to be the cornerstone of Africa's diet, a function it assumed almost four hundred years ago. Those charged with developing new types of maize, however, ought to seek a balance between producing a commercially attractive monocrop, addressing the thorny question of human disease and diet, and making use of the precious genetic diversity locked within kernels of maize themselves. After all, that potent legacy of maize—its grace—still waits to achieve full expression in the lives, economies, and soils of modern Africa.

Appendix: Tables

Table 1.1 Maize production statistics for Africa, 1997–1999, and growth rates for area, yield, and production, 1951–1999

Region/country	Average area, yield, production, 1997–1999			Growth rate of area (%/yr.)			
	Area	Yield (tons/ hectares)	Prod. ('000s of tons)	1951– 1965	1966– 1977	1978– 1987	1988– 1999
East and southern Africa	15,436	1.5	23,389	3.2	1.3	1.7	0.1
Angola	658	0.7	434	2.3	2.0	3.1	−2.9
Burundi	115	1.2	135	1.5	0.9	0.9	−1.2
Ethiopia	1,606	1.7	2,724	8.6	−1.2	1.5	8.9
Kenya	1,502	1.5	2,255	4.2	3.3	−0.1	0.2
Lesotho	134	1.0	128	−0.6	−5.2	2.2	−0.6
Madagascar	192	0.9	170	2.5	−2.1	2.2	2.7
Malawi	1,342	1.4	1,826	7.4	0.5	0.9	0.3
Mozambique	1,221	0.9	1,117	0.7	4.1	5.8	2.2
Somalia	200	0.7	141	16.7	−1.2	7.7	−0.5
South Africa	3,691	2.3	8,514	2.5	0.8	1.1	−2.1
Sudan	169	0.3	56	8.7	11.0	−4.4	8.1
Swaziland	63	1.8	115	8.8	−4.5	2.6	−3.7
Tanzania	1,785	1.3	2,362	3.7	1.8	3.6	−0.6
Uganda	615	1.2	763	2.3	5.1	−0.5	5.3
Zambia	553	1.5	818	1.0	1.1	−0.7	−4.1
Zimbabwe	1,437	1.2	1,710	5.7	1.8	3.8	2.5
West and central Africa	9,223	1.2	11,035	3.6	−0.5	7.6	1.5
Benin	607	1.2	738	1.7	−2.9	0.6	2.6
Burkina Faso	261	1.4	362	2.5	−6.3	4.3	1.1
Cameroon	392	1.5	583	7.9	2.7	14.0	7.4
Central African Rep.	94	0.9	89	18.6	8.3	−5.5	3.8
Chad	122	1.2	148	0.8	−6.6	5.9	13.5
Dem. Rep. of Congo	1,436	0.8	1,161	3.2	2.4	4.2	2.2
Ghana	683	1.5	1,015	3.5	0.9	8.6	2.8
Guinea	85	1.0	88	4.7	0.2	1.3	2.2
Ivory Coast	700	0.8	573	4.8	6.0	1.6	0.5
Mali	195	1.7	324	0.7	−2.1	13.2	4.1
Nigeria	4,111	1.3	5,419	3.3	−5.3	23.9	0.6
Senegal	61	0.9	57	8.0	−0.8	5.3	−5.4
Togo	377	1.0	384	2.1	−5.6	6.9	4.0
North Africa	1,192	5.4	6,402	−1.2	0.5	−0.6	−0.8
Egypt	864	7.1	6,164	−1.2	2.0	−0.6	<0.1
Morocco	327	0.7	237	−1.1	−1.6	−0.7	−2.8

Source: Data from Centro Internacional de Mejoridad de Maíz y Trigo (CIMMYT).

Growth rate of yield (%/yr.)				Growth rate of production (%/yr.)				As % of cereal area, 1997–1999 (average)
1951–1965	1966–1977	1978–1987	1988–1999	1951–1965	1966–1977	1978–1987	1988–1999	
1.8	2.8	−2.7	0.4	5.1	4.1	−1.0	0.5	41
1.3	−2.1	−6.0	9.3	3.6	−0.1	−2.8	6.3	76
0.4	0.7	0.4	−1.6	1.9	1.6	1.4	−2.8	56
2.1	3.3	0.8	3.4	10.7	2.1	2.3	12.3	23
1.8	2.5	3.6	−1.5	6.1	5.8	3.5	−1.3	79
0.5	2.4	−7.0	1.3	−0.1	−2.7	−4.8	0.7	71
2.4	0.5	1.4	−1.3	4.9	−1.5	3.6	1.4	14
1.0	0.9	−1.5	3.1	8.4	1.3	−0.6	3.4	89
1.5	−5.1	−7.6	12.3	2.2	−1.1	−1.8	14.5	62
0.4	0.5	6.7	−6.0	17.1	−0.7	14.4	−6.6	39
2.3	3.1	−5.7	1.6	4.8	3.8	−4.6	−0.5	74
−6.9	−0.7	−3.0	−2.9	1.8	10.3	−7.4	5.2	2
−11.0	11.6	3.3	4.4	−2.2	7.1	5.9	0.8	98
1.7	5.8	0.7	0.1	5.3	7.6	4.3	−0.5	56
1.2	2.5	−3.1	−1.1	3.5	7.5	−3.6	4.3	45
1.9	5.9	1.2	−2.0	2.9	7.0	0.4	−6.1	78
0.3	3.0	−2.6	−3.2	6.0	4.8	1.2	−0.7	75
−0.2	0.5	3.3	0.4	3.4	0.1	10.9	1.9	21
0.8	2.7	1.1	3.5	2.5	−0.2	1.7	6.1	73
1.2	1.6	−1.5	2.2	3.7	−4.6	2.8	3.3	9
−3.0	2.1	12.3	−2.1	4.8	4.8	−1.5	5.4	37
−2.0	−9.0	12.7	0.1	16.6	−0.6	7.2	3.9	62
0.9	1.4	−3.9	3.5	1.7	−5.1	2.0	17.0	6
−2.8	0.4	1.0	<0.1	0.5	2.9	5.2	2.2	69
−1.8	−1.9	0.5	1.5	1.7	−1.1	9.1	4.3	52
−2.2	−0.3	0.9	0.3	2.6	−0.1	2.2	2.5	12
6.5	−4.2	4.1	1.5	11.4	1.8	5.7	2.0	43
2.3	−0.1	−2.6	1.3	3.0	−2.2	10.6	5.5	10
0.4	2.1	−0.5	−0.2	3.7	−3.1	23.5	0.4	22
−1.5	2.8	5.1	−2.8	6.5	2.0	10.5	−8.2	5
2.5	6.3	−5.1	−0.7	4.6	0.7	1.7	3.3	52
3.1	1.5	2.6	3.4	1.9	2.0	2.0	2.6	10
2.9	0.4	2.9	3.1	1.7	2.3	2.3	3.1	32
3.9	1.3	−0.1	−5.4	2.8	−0.3	−0.8	−8.1	6

Table 3.1   Ghana: Maize area, yield, and production, 1950–1994

Year(s)	Area ('000s of hectares)	Yield (tons/hectare)	Production ('000s of tons)
1950	171	0.99	169
1963	203	0.90	183
1970–1974	421	1.07	452
1975–1979	291	1.05	304
1980–1984	441	0.76	333
1985–1989	504	1.14	573
1990–1994	590	1.49	823

Source: Robert Tripp and Kofi Marfo, "Maize Technology Development in Ghana during Economic Decline and Recovery," in Derek Byerlee and Carl Eicher, eds., *Africa's Emerging Maize Revolution* (Boulder, Colo.: Lynne Rienner, 1997), 96, cited from Department of Agriculture, Agricultural Census Office, and Ministry of Agriculture. Copyright © 1997 by Lynne Rienner Publishers, Inc. Reprinted with permission of the publisher.

Table 4.1 Production of cereals in regions of Italian East Africa, 1938
(percentages)

Cereal	Eritrea	Amara	Shawa	Harer	Galla and Sidamo	Somalia
Sorghum	15.96	24.87	10.31	43.82	19.05	80.00
Teff	21.28	39.80	15.46	5.58	19.05	—
Barley	34.57	22.39	41.25	27.89	6.67	—
Wheat	18.62	2.99	20.62	9.96	1.90	—
Maize	3.19	7.46	10.30	12.75	47.62	20.00
Millet	6.38	2.49	2.06	—	5.71	—

Source: Statistics from Rafaele Ciferri and Enrico Bartolozzi, "La produzione
cerealicola dell'Africa Orientale Italiana nel 1938," *L'Agricoltura Coloniale*
(Florence) 19, nos. 11–12 (1940): 6–7.

Table 5.1   White maize production in Africa

Country/ Region	1979–1981 average ('000s of metric tons)	1989–1991 average ('000s of metric tons)	1996 ('000s of metric tons)	Production growth (%) 1979–1981 1989–1991	Per capita growth (%) 1979–1981 1989–1991	White maize as % of total production
North Africa						
Egypt	3,001	4,576	5,318	4.3	1.7	95
West Africa						
Benin	274	401	479	3.9	1.6	95
Ghana	342	660	907	6.8	3.3	90
Ivory Coast	246	348	391	3.5	−0.3	70
Nigeria	539	1,759	n.a.	12.6	9.0	90
Togo	134	241	373	6.0	2.9	90
Central Africa						
Angola	303	228	398	−2.8	−5.4	100
Cameroon	397	382	622	−0.4	−3.2	95
Zaire	574	949	1,044	5.2	1.8	95
East Africa						
Ethiopia	1,224	1,771	3,260	3.8	1.2	100
Kenya	1,714	2,420	2,079	3.5	0.0	100
Somalia	120	238	142	7.1	4.4	100
Tanzania	1,762	2,634	2,314	4.1	0.7	100
Uganda	361	598	800	5.2	2.1	100
Southern Africa						
Lesotho	112	139	199	2.2	−0.5	100
Malawi	1,275	1,481	1,793	1.5	−2.8	100
Mozambique	383	370	947	−0.3	−2.0	100
South Africa	4,882	4,909	5,836	0.1	−2.4	45
Swaziland	85	122	136	3.7	0.8	100
Zambia	941	1,345	1,410	3.6	0.1	100
Zimbabwe	1,738	1,766	2,375	0.2	−3.1	95
Africa (total)	20,407	22,761	30,823	3.7	0.7	

Source: CIMMYT and FAO, *White Maize: A Traditional Food Grain in Developing Countries* (Rome: CIMMYT and FAO, 1997).

Table 6.1 Maize production in colonial Africa, 1950 and 1960

Country (colony)	Production ('000s of metric tons) ca. 1950	1960
British Empire		
Bechuanaland	1	3
Gambia	7	n.a.
Gold Coast (Ghana)	200	205
Kenya	574	1,074
Malawi	274	n.a.
Nigeria	543	963
Northern Rhodesia	180	n.a.
Sierra Leone	8	8
Southern Rhodesia	250	1,014
Tanganyika	393	538
Uganda	139	225
French empire		
Cameroon	98	177
Central African Republic	16	n.a.
Chad	2	9
Congo (Brazzaville)	3	1
Dahomey	155	193
Gabon	1	1
Guinea	62	58
Ivory Coast	63	132
Malagasy	48	78
Mali	65	60
Mauritania	5	3
Mauritius	4	1
Niger	2	3
Réunion	8	16
Rwanda-Burundi	101	169
Senegal	14	29
Togo	62	52
Upper Volta	86	55
Portuguese empire		
Angola	320	326
Mozambique	262	n.a.
Other		
Congo (Belgian)	313	276
Ethiopia	160	167
Liberia	20	11
Sudan	15	35

Source: Updated and adapted from Marvin P. Miracle, *Maize in Tropical Africa* (Madison: University of Wisconsin Press, 1966), 83.

Table 7.1 Zimbabwe maize sales to grain marketing board by sectors, 1979–1990

| Season | Large-scale commercial | Small-scale farm sector | | | Total (tons) | Grand total | Imports |
		Small-scale commercial	Communal lands	Resettlement land			
1979–80	473,736 (92.4)[a]	16,079 (3.1)	23,105 (4.5)	—	39,184 (7.6)	512,920	
1980–81	728,532 (89.3)	21,053 (2.6)	66,565 (8.2)	—	87,618 (10.7)	816,150	94,916
1981–82	1,650,574 (82.0)	72,786 (3.6)	290,488 (14.4)	—	363,274 (18.0)	2,013,848	
1982–83	1,021,892 (73.4)	52,591 (3.8)	317,884 (22.8)	—	370,475 (26.6)	1,392,367	
1983–84	464,486 (75.3)	15,453 (2.5)	137,243 (22.2)	—	152,696 (24.7)	617,182	
1984–85	551,612 (58.6)	55,333 (5.9)	335,130 (35.6)	—	390,463 (41.1)	942,075	
1985–86	1,008,971 (57.6)	68,431 (3.8)	666,604 (38.0)	62,604 (3.5)	741,299 (42.4)	1,750,270	268,935
1986–87	911,945 (57.2)	—	—	—	682,429 (42.8)	1,594,374	
1987–88	246,735 (61.3)	—	—	—	155,755 (38.7)	402,490	
1988–89	440,773 (36.9)	—	—	—	754,227 (63.1)	1,195,000	
1989–90	491,382 (49.3)	—	—	—	506,001 (50.7)	997,383	

a. Figures in parentheses are percentage of total marketed production.

Source: Adapted from Kingston Mashingaidze, "Maize Research and Development," in Mandivamba Rukuni and Carl Eicher, eds., *Zimbabwe's Agricultural Revolution* (Harare: University of Zimbabwe, 1997), 209, with corrections from Grain Marketing Board.

Table 7.2 Maize modern varieties in four African countries, 1990–1999

Country	1990		1996		1999	
	% of hybrids	% of all modern varieties[a]	% of hybrids	% of all modern varieties	% of hybrids	% of all modern varieties
Kenya	62	70	64	73	85	87
Malawi	11	14	33	37	39	43
Zambia	72	77	22	23	62	65
Zimbabwe	96	96	91	96	91	100

a. *Modern varieties* indicates hybrids and improved open-pollinated seed.

Source: Melinda Smale and Thomas Jayne, *Maize in Eastern and Southern Africa: Seeds of Success in Retrospect*, EPTD Discussion Paper no. 97, Environmental and Technical Division, International Food Policy Research Institute, January 2003. Used by permission.

Table 8.1 Cereals and maize: Area under cultivation in Ethiopia ('000s of hectares), and production ('000s of tons), 1989–1998

Year	Cereals[a]		Total		Maize					
	Area	Prod.	Area	Prod.	Area	Prod.	% of total area	% of total prod.	% of cereal area	% of cereal prod.
1989	4,848	56,859	5,577	63,200	1,021	16,887	14.6	26.7	21.0	29.7
1990	4,915	60,888	5,706	67,559	1,278	20,556	22.4	30.4	26.0	33.8
1991	4,295	55,779	5,154	67,622	1,121	13,479	21.8	19.9	26.1	24.2
1992	4,263	49,290	5,114	65,905	986	15,106	19.3	27.0	23.1	30.6
1993	3,954	51,487	4,857	57,950	809	13,915	16.7	24.0	20.5	27.0
1994	5,287	51,052	7,158	57,001	1,208	13,379	16.9	23.5	22.8	26.2
1995	6,449	58,484	7,681	67,428	1,418	13,637	18.5	20.0	22.0	23.3
1996	7,670	92,654	9,026	102,877	1,851	31,053	20.5	30.0	24.1	33.5
1997	6,689	86,293	8,011	91,152	1,317	25,320	16.4	27.8	19.7	29.3
1998	6,313	71,974	8,186	80,662	1,449	23,443	17.7	29.1	23.0	32.6

a. Cereals category includes maize, teff, barley, eleusine, wheat, and sorghum.
Source: Data from Tesfaye Tsegaye and Alemu Hailye, Adoption of Improved Maize Technologies and Inorganic Fertilizer in Northwestern Ethiopia, Research Report no. 40 (Addis Ababa: Ethiopian Agricultural Research Organization, 2001).

Table 8.2   Maize and cereal cultivation in Burie district, 1993–2002

Year	Total area (hectares) planted to maize	Total cereal area planted (% of total)
1993	11,331	63,795 (17.7)
1994	11,933	59,114 (20.1)
1995	12,882	61,012 (21.1)
1996	13,268	62,921 (21.0)
1997	13,727	66,217 (20.7)
1998	17,642	73,834 (23.8)
1999	19,460	77,078 (25.2)
2000	26,902	79,143 (33.9)
2001	26,758	78,140 (35.9)
2002	19,929	74,505 (26.7)[a]

a. The year 2002 was one of major failure of the main rains; farmers shifted maize fields to short-season wheat, and maize yield fell by half.

Source: Data supplied by Ato Kurabachew Mezgebu and Ato Desta Yusufu, Burie and Wemberma District Agriculture Bureau.

Notes

⁂

1. Africa and the World Ecology of Maize

1. For a full description of the concept of legibility, see James Scott, *Seeing Like a State: How Certain Schemes to Improve the Human Condition Have Failed* (New Haven, Conn.: Yale University Press, 1998), 2–3.

2. Worldwide, the figure for human consumption is 34 percent (the remainder being fed to livestock). Derek Byerlee and Carl Eicher, *Africa's Emerging Maize Revolution* (Boulder, Colo.: Lynn Rienner, 1997), 16; Christopher Dowswell, R. L. Paliswal, and Ronald P. Cantrell, *Maize in the Third World* (Boulder, Colo.: Westview, 1996), 27.

3. In modern maize breeding the breeders cover the tassels with brown paper bags to control and then to collect the pollen used in cross-pollinating particular plants.

4. For a summary of this debate, see Betty Fussell, *The Story of Corn: The Myths and History, the Culture and Agriculture, the Art and Science of America's Quintessential Crop* (New York: Knopf, 1992), 76–86; Paul Mangelsdorf, *Corn: Its Origin, Evolution, and Improvement* (Cambridge, Mass.: Harvard University Press, 1974), 11–26, 48–49. The annual teosinte *(Zea mexicana)* is a likely ancestor, and the closest relative, of maize. Like *Zea mays* it has ten pairs of chromosomes, which are cytogenically similar to those of maize, and the plants hybridized freely, producing fertile progeny. Both maize and teosinte have male flowers in their tassels and female flowers on the lateral branch. The species differ primarily in their female organs. Maize depends on humans to separate and disperse the seed from the highly protected ear, while teosinte is self-sowing and shatters at maturity. See Dowswell, Paliswall, and Cantrell, *Maize in the Third World*, 17.

5. It has been pointed out to me by my friend Cassandra Moseley that human husbandry has also erased part of maize's genetic memory.

6. For technical descriptions of hybrid types and methods see Dowswell, Paliswall, and Cantrell, *Maize in the Third World,* 141.

7. I am grateful to Vincent Knapp for information on the chemistry of vegetables. For maize chemistry, see "Quality Protein Maize," *CIMMYT Today* 1 (1975): 1–12; Hugo Cordova, "Quality Protein Maize: Improved Nutrition and Livelihoods for the Poor," in Maize Program, *Maize Research Highlights, 1999–2000* (Mexico City: CIMMYT, 2001), 27–31.

8. Centro Internacional de Mejoridad de Maíz y Trigo (henceforth CIMMYT) and Food and Agriculture Organization of the United Nations (henceforth FAO), *White Maize: A Traditional Food Grain in Developing Countries* (Rome: CIMMYT and FAO, 1997), 7. Also see Melinda Smale and Thomas Jayne, *Maize in Eastern and Southern Africa: "Seeds" of Success in Retrospect,* EPTD Discussion Paper no. 97 (Washington, D.C.: Environmental and Technical Division, International Food Policy Research Initiative, 2003), 11.

9. Elon Gilbert, "The Meaning of the Maize Revolution in Sub-Saharan Africa: Seeking Guidance from Past Impacts," Overseas Development Administration Network Paper no. 55, 19; Melinda Smale, " 'Maize is Life': Malawi's Delayed Green Revolution," *World Development* 23, no. 5 (1995): 820. Smale cites this expression from Pauline Peters, who originally quoted it from speakers in the Zomba area of Malawi.

10. Dowswell, Paliswall, and Cantrell, *Maize in the Third World,* 8.

11. Most Ethiopians are shocked by this fact. I recently asked an Ethiopian friend from a rural background to guess Ethiopia's major crop. She guessed wrong five times and never ventured maize as a possible answer. West African maize has a lesser but growing role in the region's total calories. In Ghana maize production has tripled since 1950 and has become that country's dominant cereal. Robert Tripp and Kofi Marfo, "Maize Technology Development in Ghana during Economic Decline and Recovery," in Derek Byerlee and Carl Eicher, *Africa's Emerging Maize Revolution,* 95.

12. Byerlee and Eicher, "Accelerating Maize Production: Synthesis," in ibid., 249–250.

13. Norman Borlaug, "Linking Technology and Policy," in Steven Breth, ed., *Overcoming Rural Poverty in Africa* (Geneva: Center for Applied Studies in International Negotiations, 1997), 140–142. Borlaug has championed the position that off-the-shelf technology and crop varieties already exist to transform world food supply, but that they must be linked with strenuous national efforts at extension.

14. P. L. Pingali, ed., *CIMMYT 1999–2000 World Maize Facts and Trends. Meeting World Maize Needs: Technological Opportunities and Priorities for the Public Sector* (Mexico City: CIMMYT, 2001), 2–3.

15. Pingali, *1999–2000 World Maize Facts,* 50, 52.

16. E. J. Wellhausen, "Recent Developments in Maize Breeding in the Tropics," in David B. Walden, ed., *Maize Breeding and Genetics* (New York: Wiley and Sons, 1978), 59.

17. Paul Heisey and Gregory O. Edmeades, eds., *World Maize Facts and Trends, 1997–1998* (Mexico City: CIMMYT, 1999), 37–38.

18. Jonathan Kingdon, *Island Africa: The Evolution of Africa's Rare Animals and Plants* (Princeton, N.J.: Princeton University Press, 1989), 146–148.

19. Pingali, *1999–2000 World Maize Facts,* 6–7.

20. Peter Sutcliffe, "Soil Conservation and Land Tenure in Highland Ethiopia," *Ethiopian Journal of Development Research* 17, no. 1 (1995): 63–87.

21. Eugene Rasmussan, "Global Climatic Change and Variability: Effects of Drought and Desertification in Africa," in Michael Glantz, ed., *Drought and Hunger in Africa: Denying Famine a Future* (Cambridge: Cambridge University Press, 1987), 3–22.

22. See, for example, Peter J. Lamb, "Large Scale Tropical Atlantic Surface Circulation Patterns Associated with Sub-Saharan Weather Anomalies," *Tellus* 39 (1978): 240–251.

23. Within the past few millennia, however, there have been wetter periods. They included one in which the Sahara was a pastoral grassland that supported cattle, game, and human settlement. See James C. McCann, *Green Land, Brown Land, Black Land: An Environmental History of Africa* (Portsmouth, N.H.: Heinemann, 1999), 16–18.

24. New efforts have also included research by CIMMYT in Harare, Zimbabwe, to develop drought tolerance. The development of Katumane varieties in Kenya and the SR200 series in Southern Rhodesia (Zimbabwe) in the 1970s were major breakthroughs in adjusting for marginal lands occupied by African smallholders. See John Gerhart, *The Diffusion of Hybrid Maize in Western Kenya* (Mexico City: CIMMYT, 1975); William A. Masters, *Government and Agriculture in Zimbabwe* (Westport, Conn.: Greenwood Press, 1994), 50–51. See also Chapter 8.

25. Pingali, *1999–2000 World Maize Facts and Trends,* 7.

26. Heisey and Edmeades, *World Maize Facts and Trends, 1997–1998,* 5–6.

27. Michael Bratton, "Drought, Food, and the Social Organization of

Small Farmers in Zimbabwe," in Glantz, *Drought and Hunger in Africa,* 213–244. See also Amartya Sen, *Poverty and Famines: An Essay in Entitlements and Deprivation* (Oxford: Oxford University Press, 1981).

2. Naming the Stranger: Maize's Journey to Africa

1. See Alfred Crosby, *The Columbian Exchange: Biological and Cultural Consequences of 1492* (Westport, Conn.: Greenwood Press, 1972). A few scholars have claimed, citing early Portuguese references to *milho zaburro* on the West African coast, that maize in the Old World antedated 1492.

2. M. D. W. Jeffreys, "The History of Maize in Africa," *South African Journal of Science* (March 1954): 198. In this quotation the common mistranslation of the Portuguese *milho zaburro* as "maize" (as opposed to sorghum, an indigenous African grain) is not a factor. The reference to *mehiz* and to the grain's resemblance to chickpeas points to maize as the cereal described.

3. Ibid., 121. Jeffreys cites a translation by Lains e Silva. See also Frank Willett, "The Introduction of Maize into West Africa: An Assessment of Recent Evidence," *Africa* 32, no. 1 (1962): 11. Robert Harms tells me that the French slaving ship *Diligent* called in at São Tomé in December–January 1731–1732, where it took on cassava flour and "une Demy gamelle De mil" (a half barrel of maize / sorghum) as food for the middle passage. The latter may have been maize rather than millet, since maize is elsewhere described as a major provision for such vessels calling at São Tomé. It is not clear why the *Diligent* did not take on a larger consignment of maize, though perhaps the harvest was delayed and supplies were low.

4. O. Dapper, *Nankeurige Beschrijvinge der Afrikaensche Gewestern* (Amsterdam, 1668), 463. Quoted in Helma Pasch, "Zur Geschichte der Verbreitung des Maizes in Afrika," *Sprache und Geschichte in Afrika 5* (1983): 189.

5. Dominique Juhé-Beaulaton, "La diffusion du maïs sur les Côtes de l'Or et des Esclaves aux XVII et XVIII siècles," *Review Français d'Histoire d'Outre-Mer* 77 (1990): 188–190.

6. Roger Blench, Kay Williamson, and Bruce Connell, "The Diffusion of Maize in Nigeria: A Historical and Linguistic Investigation," *Sprache und Geschichte in Afrika* 15 (1994): 109–114. For assertions about the Portuguese role see, for example, Marvin P. Miracle, *Maize in Tropical Africa* (Madison: University of Wisconsin Press, 1966), which cites no evidence but seems to be the source cited by most writers.

7. Jeffreys, "The History of Maize in Africa," 197–200, and M. D. W. Jeffreys, "The Origin of the Portuguese Word *Zaburro* as Their Name for Maize," *Bulletin de L'Institut Français de l'Afrique Noire,* series B, *Sciences Humaines* 19, nos. 1–2 (January–April 1957): 111–136. For Bascom's complaint and the full body of evidence from Nigeria, see Blench, Williamson, and Connell, "The Diffusion of Maize in Nigeria," 9–46.

8. A. J. H. Goodwin, "The Origin of Maize," *South African Archaeological Bulletin* 8 (1953): 13.

9. The primary maize cultivated for polenta in Veneto region of Italy until 1952 was a red flint called Marano. I am grateful to Emma Meggiolaro and Armando DeGuio for this information. Meggiolaro cultivated maize as a young girl in the Veneto district of near Vicenza. The colloquial Venetian term for maize is *sorgo rosso* (red sorghum).

10. Joseph Burtt-Davy, *Maize: Its History, Cultivation, Handling, and Uses. With Special Reference to South Africa* (London: Longmans, Green, 1914), 316. There was a white-grain version of this flint found on black South African farms. For further description of Brazilian Cateto flint maize, see also Christopher Dowswell, Christopher, R. L. Paliswal, and Ronald P. Cantrell, *Maize in the Third World* (Westport, Conn., 1996), 17, 106–107; E. J. Wellhausen, "Recent Developments in Maize Breeding in the Tropics," in David B. Walden, ed., *Maize Breeding and Genetics* (New York: Wiley and Sons, 1978), 61.

11. Juhé-Beaulaton, "La diffusion," 187, states that the seventeenth-century inhabitants of Aquapim consumed maize exclusively fresh or roasted on the cob.

12. Portères argues that floury maize types came via the West African coast, and flint types via the Nile Valley. Miracle suggests that flints may also have come from the coast but may have found less favor among farmers, especially women, who preferred the soft starches better suited to hand-milling. Most current evidence indicates that African women prefer flinty types for hand-milling and storage. Current "local" maize varieties in Ghana are semident floury types. Interview with S. Twumasi-Afriyie, Kumasi, July 1997. See Roland Portères, "L'introduction du maïs en Afrique," *Journal de l'Agriculture Tropicale et de Botanique Appliquée* 2, nos. 5–6 (May-June 1955): 224–226; Marvin P. Miracle, "Interpretation of Evidence on the Introduction of Maize into West Africa," *Africa* 33, no. 2 (1963): 133. Burtt-Davy mentions a Brazilian Flour Corn that resembles a Zulu bread maize known as *u'hlanza-gazaan* (white man's grain). See Burtt-Davy, *Maize,* 322–323.

13. Dowswell, Paliswal, and Cantrell, *Maize in the Third World,* 48–49.

14. See James C. McCann, *People of the Plow: An Agricultural History of Ethiopia* (Madison: University of Wisconsin Press, 1995), 56. For an earlier date for pod maize, see Burtt-Davy, *Maize,* 13. His claim is, however, doubtful.

15. Henry Stanley, *Through the Dark Continent* (New York: Harper Brothers, 1878), 1:439.

16. G. Schweinfurth, ed., *Emin Pasha in Central Africa* (London: Phillip and Son, 1888), 418.

17. R. S. Rattray, *Ashanti* (Oxford: Oxford University Press, 1923), 47. Also see Ivor Wilks, *Forests of Gold: Essays on the Akan and the Kingdom of Asante* (Athens: Ohio University Press, 1993), 66–72.

18. J. G. Christaller, ed., *Three Thousand and Six Hundred Ghanaian Proverbs* (New York: Mellen Press, 1990), 150. I am indebted to the work of James D. LaFleur and to the advice of Maxwell Amoh and Edward Kissi for this citation.

19. A. Babalola, "Folk Tales from Yoruba Land," *West Africa Review* 23 (1952): 14. Willett, "The Introduction of Maize into West Africa," 9, ascribes this tradition to the early nineteenth century.

20. Wande Abimbola, *Ifá: An Exposition of Ifá Literary Corpus* (New York: n.p., 1997), 221–222.

21. Quoted in Joseph Burtt-Davy, *Maize,* 1.

22. Robert Soper, "Roulette Decoration on African Pottery: Technical Considerations, Dating, and Distributions," *African Archaeological Review* 3 (1985): 31, 44. For a report on the Fon tradition, see Juhé-Beaulaton, "La diffusion," 197.

23. A. C. A. Wright, "Maize Names as Indicators of Economic Contacts," *Uganda Journal* 13, no. 1 (1949): 64. Wright mentions a Portuguese manuscript of 1634 that refers to maize production on Pemba Island. Steven Feierman and Christopher Ehret, personal communication. Also see Helma Pasch, "Zur Geschichte der Verbreitung des Maises in Afrika," *Sprache und Geschichte in Afrika* 5 (1983): 196–197.

24. My thanks to Edward Kissi, Clark University, for this information.

25. For many other West African languages, the evidence is not at all clear. For a comprehensive list of West African maize terms, see the exhaustive list compiled in 1854 by Sigismund Koelle, *Polyglotta Africana,* ed. P. E. H. Hair and David Dalby (Graz, Austria: Akademische Druk-U., 1963), 104–105, 112–113. Also see Christine Chiang, "Determining the Origins of Maize in West Africa Using Ethnolinguistic Evidence," unpublished paper, University of California at Los Angeles, 1997, 4. These names suggest possible patterns of adoption, though any conclusions can only be

speculative and must await a more detailed, systematic linguistic survey. In India the Hindi word *makka,* or *makka jouar,* implies an introduction from Arabia and the eastern Mediterranean. In each of these cases it seems likely that maize returned with *hajji,* whose pilgrimages provided both intellectual and agronomic cross-fertilization.

26. Burtt-Davy, *Maize,* 13–14; 18–24.

27. Ibid., 13. I am grateful to Sandra Sanneh of the Department of African Languages at Yale University for help on South African languages.

28. David Gough, *A Reconsideration of Some Southern Bantu Maize Terms,* Essays in Bantu Language Studies, Working Paper no. 7 (Grahamstown, South Africa: Department of African Languages, Rhodes University, 1981), 2–6. Gough also points out that the old Zulu term *vila* (sorghum) is a likely source for the term for maize, *mbila.*

29. Christopher Ehret, "Agricultural History in Central and Southern Africa, ca. 1000 B.C. to A.D. 500," *Transafrican Journal of History* 4 (1974): 5.

30. This section on Bantu glosses relies heavily on the work of Christopher Ehret, who generously commented on my Bantu maize word list with insights from his great experience in working with that language family as well as in Cushitic and Semitic languages in eastern and southern Africa.

31. Ehret, "Agricultural History in Central and Southern Africa," 5–7.

3. Maize's Invention in West Africa

1. See Jared Diamond, *Guns, Germs, and Steel: The Fates of Human Societies* (New York: Norton, 1997). Christopher Ehret disagrees and posits African innovation in sorghum as an example of the continent's participation in human agricultural invention. See Christopher Ehret, *An African Classical Age: Eastern and Southern Africa in World History, 1000 B.C. to A.D. 400* (Charlottesville: University of Virginia Press, 1998).

2. For McNeill's comments on maize, see William H. McNeill, "American Food Crops in the Old World," in Herman J. Viola and Carolyn Margolis, eds., *Seeds of Change: A Quincentennial Commemorative* (Washington, D.C.: Smithsonian Institution Press, 1991), 43–59.

3. Alfred Crosby, *The Columbian Exchange: Biological and Cultural Consequences of 1492* (Westport, Conn.: Greenwood Press, 1972), 168.

4. James Fairhead and Melissa Leach, *Reframing Deforestation: Global Analyses and Local Realities, Studies in West Africa* (London: Routledge, 1998).

5. This case draws on a range of historical and geographic research

summarized in James C. McCann, *Green Land, Brown Land, Black Land: An Environmental History of Africa* (Portsmouth, N.H.: Heinemann, 1999), chap. 4.

6. For evidence of the historical geography of cereal on the West African coast, see Dominique Juhé-Beaulaton, "La diffusion du maïs sur les Côtes de l'Or et des Esclaves aux XVII et XVIII siècles," *Review Français d'Histoire d'Outre-Mer* 77 (1990): 185–186, 190–191.

7. For his views on the ecology of yams, I am grateful to Edward Ayensu, former ecologist at the Smithsonian Institution.

8. Ivor Wilks, *Forests of Gold: Essays on the Akan and the Kingdom of Asante* (Athens: Ohio University Press, 1993), 58. Wilks cites J. Dupuis, *Journal of a Residence in Ashantee* (London: Cass, 1966), 65–66.

9. Wilks, *Forests of Gold*, 58.

10. For a list of terms for maize in West African languages, see Sigismund Koelle, *Polyglotta Africana*, P. E. H. Hair and David Dalby, eds. (Graz, Austria, 1963), 104–105, 112–115.

11. I am grateful to historian James LaFleur for this reference from his work with Dutch documents on the early Gold Coast. Speakers of Ghana's Twi language argue for the gloss "foreign" for the maize name *aburo* (or *aburoo*). This confusion probably derives from a widely adopted folk etymology that associates maize with both sorghum and foreign origin.

12. Kojo Amanor, *The New Frontier: Farmer's Response to Land Degradation. A West African Study* (London: Zed Books, 1994), 173–178. For the arguments that this is a historical process and not merely recent adaptation, see James Fairhead and Melissa Leach, *Misreading the Landscape: Society and Ecology in a Forest-Savanna Mosaic* (Cambridge: Cambridge University Press, 1996), 289; Fairhead and Leach, *Reframing Deforestation*, 60–93; and Wilks, *Forests of Gold*, 56–63.

13. Amanor, *The New Frontier*, 175.

14. Juhé-Beaulaton, "La diffusion," 185–186; 190–191.

15. Christopher Dowswell, R. L. Paliswal, and Ronald P. Cantrell, *Maize in the Third World* (Boulder, Colo., 1996), 106–107.

16. Robert Tripp and Kofi Marfo, "Maize Technology Development in Ghana during Economic Decline and Recovery," in Derek Byerlee and Carl Eicher, eds., *Africa's Emerging Maize Revolution* (Boulder, Colo.: Lynne Rienner, 1997), 96.

17. For more on the swollen shoot outbreak, see ibid., 96.

18. Interview with S. Twumasi-Afriyie, Crop Research Institute, Kumasi, June 1997.

19. Tripp and Marfo, "Maize Technology," 95.

20. For this philosophy and program goals for QPM, see Centro Internacional de Mejoridad de Maíz y Trigo (CIMMYT) and Food and Agriculture Organization of the United Nations (FAO), *The Quality Protein Maize Revolution: Improved Nutrition and Livelihoods for the Poor* (Mexico City: CIMMYT, n.d.).

21. See the list of words for maize in West African languages, Koelle, *Polyglotta Africana,* passim. Wande Abimbola, a professor who is a widely acknowledged master of the Yoruba language and oral culture, has told me that he knows of no direct gloss for *agbado* besides "maize." This may be a reflection of the early arrival of maize in the Bight of Benin.

22. See Roger Blench, Kay Williamson, and Bruce Connell, "The Diffusion of Maize in Nigeria: A Historical and Linguistic Investigation," *Sprache und Geschichte in Afrika* 15 (1994), 19.

23. Most scholars have accepted Marvin Miracle's assumptions about the Portuguese origins of West African and southern African maize. Miracle, *Maize in Tropical Africa* (Madison: University of Wisconsin Press, 1966), 87–100, 285.

24. C. L. M. Van Eijnatten, *Towards the Improvement of Maize in Nigeria* (Wageningen, Netherlands: Veenman and Zonen N. V., 1965).

25. J. J. Smith, Georg Weber, M. V. Manyong, and M. A. B. Fakorede, "Fostering Sustainable Increases in Maize Productivity in Nigeria," in Byerlee and Eicher, *Africa's Emerging Maize Revolution,* 111–113.

26. Ibid., 117.

4. Seeds of Subversion in Two Peasant Empires

1. The links in material culture and ideology between Ethiopia and the Mediterranean world have been recently confirmed by the joint archaeological project at Aksum involving Boston University, the University of Naples, and the Ethiopian Ministry of Culture. See Kathryn Bard, ed., *The Environmental History and Human Ecology of Northern Ethiopia in the Late Holocene: Preliminary Results of Multidisciplinary Project* (Naples: Università Orientale, 1997).

2. Carlo Ginzburg, *The Cheese and the Worms: The Cosmos of a Sixteenth-Century Miller* (Baltimore, Md.: Johns Hopkins University Press, 1982), 119–120.

3. A few scholars have claimed that maize in the Old World antedated 1492, citing early Portuguese references to *milho zaburro* on the West African coast. See M. D. W. Jeffreys, "The History of Maize in Africa," *South African Journal of Science* (March 1954): 197–200; M. D. W. Jeffreys, "The

Origin of the Portuguese Word *Zaburro* as Their Name for Maize," *Bulletin de l'Institut Français de l'Afrique Noire,* series B, *Sciences Humaines* 19, nos. 1–2 (January–April 1957): 111–136. For a refutation of this thesis, see Paul C. Mangelsdorf, *Corn: Its Origin, Evolution, and Improvement* (Cambridge, Mass.: Harvard University Press, 1974), 201–206.

4. Ginzburg, *The Cheese and the Worms,* 82–86. Ginzburg presents the miller Menocchio's view of the news, as it filtered down to the village, of a new world as an intimation of a new spiritual and social realm yet to unfold, rather than simply a reference to new lands for commercial exploitation.

5. Jan DeVries, *The Economy of Europe in an Age of Crisis, 1600–1750* (Cambridge: Cambridge University Press, 1976), 53–55; Ginzburg, *The Cheese and the Worms,* 15; Peter Musgrave, *Land and Economy in Baroque Italy: Valpolicella, 1630–1797* (Leicester, England: Leicester University Press, 1992), 3–5.

6. See Vincent Knapp, "What Europeans Ate in Agricultural Times: Eighteenth-Century Levels of Food Consumption and Nutrition," paper presented to the Seminar in Agrarian Studies, Yale University, 22 January, 1999, 2–6. The harder durum wheat was used for pasta, the southern staple. For the importation of Baltic rye, see Musgrave, *Land and Economy,* 18.

7. Agronomically, rice could occupy about only two hundred thousand hectares. Interview with Marco Bertolini and Alberto Vederio, Istituto per la Cerealicoltura, Bergamo, Italy, June 1999.

8. Domenico Sella, *Crisis and Continuity: The Economy of Spanish Lombardy in the Seventeenth Century* (Cambridge: Cambridge University Press, 1979), 6–7.

9. For the most thorough description of water management in northern Italy, see Salvatore Ciriacono, "Investimenti capitalistici e colture irrigue la congiuntura agricola nella terraferma veneta (secoli XVI–XVII)," *Atti del Convegno Venezia e la Terraferma attraverso le Relazione dei Rettori,* 1980, 123–128. Ciriacono attributes early water management in Lombardy to Iberian influences and local innovation. See Salvatore Ciriacono, "Introduction," in Salvatore Ciriacono, ed., *Land Drainage and Irrigation* (Aldershot, England: Ashgate Publishing, 1998), xx.

10. John McIntire, Danile Bourzat, and Prabhu Pingali, *Crop-Livestock Interactions in Sub-Saharan Africa* (Washington, D.C.: World Bank, 1990).

11. Around that time as well, Giovanni Lamo, a Venetian, sent some seeds to the duke of Florence, inviting him to plant them. Luigi Messedaglia, *Il mais e la vita rurale Italiana: Saggio di storia agraria* [Maize and Ital-

ian Rural Life: An Essay on Agrarian History] (Piacenza, Italy: Federazione Italiana dei Consorzi Agrari, 1927), 178.

12. See Henry Lowood, "The New World and the European Catalog of Nature," in Karen Ordahl Kupperman, ed., *America in European Consciousness* (Chapel Hill: University of North Carolina Press, 1995), 295–323.

13. Michele Fassina, "Il mais nel Veneto nel cinquecento: Testimonianze iconografiche e prime esperienze colturali" [Maize in the Veneto in the Sixteenth Century: Iconographic Testimonies and First Cultural Experience], in *L'impatto della scoperta dell'America nella cultura veneziana* (Rome: Bulzone Editore, n.d.), 86. The plants depicted in the *Tentazione de Adamo* bas-relief may actually be sorghum, rather than maize, given the early date (1515) and the difficulty of distinguishing between the two crops in their early growth.

14. Messedaglia, *Il mais*, 29–39. Not in Messedaglia's list is *sorgo rosso*, a common term for maize in current Venetian dialect. See Michele Fassina, "Elementi ed aspetti della presenza del mais nel Vicentino: Con particolare riferimento a Lisiera e alla zona attraversata dal fiume Tesina" [Elements and Aspects of the Presence of Maize in the Vicentino. With Particular Reference to Lisiera and to the Zone across the Tesina River], in *Lisiera: Immagini, documenti per la storia e cultura di una comunità veneta—strutture, congiunture, episodi* (Lisiera, Italy: Edizioni parrocchia de Lisiera, 1981), 314–316.

15. Christopher Dowswell, R. L. Paliswal, and Ronald P. Cantrell, *Maize in the Third World* (Boulder, Colo.: Westview, 1996), 18. Dowswell suggests that the seed was yellow Caribbean flint, though other evidence suggests the variety was a red flint. See also Roland Portères, "L'introduction du maïs en Afrique," *Journal de l'Agriculture Tropicale et de Botanique Appliquée* 2, nos. 5–6 (May–June 1955): 221–231.

16. Literally, "red sorghum," a name for maize still used, as mentioned earlier, in the Venetian dialect and a variety that also became the first maize appearing in Egypt under local Arabic names signifying "red in color" (*durra hamra*—lit., "red sorghum"). Interview with Emma Meggiolaro, 29 June 1999, Montecchia.

17. See Fassina, "Elementi ed aspetti," 321.

18. For examples of Venetian land surveys, see Silvestro Camerini, *La possidenza borghese in Transpadana* [Middle-Class Landowners in Transpadana] (Padua: n.p., 1991), 119–194.

19. For a description of stall-feeding of cattle and draft animals, see Piero Rizzolatti, "La stalla e il governo degli animali" [The Stall and the Manage-

ment of Animals], in Giovan Battista Pellegrini, ed., *I lavori dei contadini* (Vicenza, Italy: Neri Pozza, 1997), 333–384. A beautiful illustration of the role of the stall on northern Italian peasant estates can be found in the film *The Tree of the Wooden Clogs*.

20. Michele Fassina, "L'introduzione della coltura del mais nelle campagne venete" [Introduction to the Cultivation of Corn in the Venetian Countryside], *Società e Storia* 15 (1982): 36, 39.

21. Quoted in Dava Sobel, *Galileo's Daughter: A Historical Memoir of Science, Faith, and Love* (New York: Penguin, 2000), 52–53.

22. Fassina, "Il mais nel Veneto nel cinquecento," 91.

23. Michele Fassina, "Aspetti economici e sociali in una grande azienda agricola polesana nel corso del XVII secolo" [Economic and Social Aspects of a Large Polesine Farm in the Seventeenth Century], *Uomini, Terra e Acque, Atti del XIV Convegno di Studi Sorici organizzato in collaborazione con l'Accademia del Concordi* (Rovigo, Italy: Associazione Culturale Minelliana, 1990), 229.

24. Joseph Burtt-Davy reports that a certain Caspar Bauchin claims to have seen a pod maize (*Zea mays L. var. tunicata St. Hil.*) in Ethiopia under the name of *manigette* in Ethiopia in 1623. Merid Wolde Aregay doubts that reference and has told me that his perusal of Portuguese and missionary sources for the seventeenth century revealed no references to maize. See Joseph Burtt-Davy, *Maize: Its History, Cultivation, Handling, and Uses* (London: Longmans, 1914), 13.

25. For a general summary of this period in Ethiopia, see Harold G. Marcus, *A History of Ethiopia* (Berkeley: University of California Press, 1994), 39–41. For similar, more lasting changes in ideology and social structure in the Nile Valley Sennar kingdom, see Jay Spaulding, *The Kingdom of Sennar* (East Lansing: Michigan State University Press, 1982).

26. Personal conversation with Merid, Addis Ababa, 5 June 2003. I am grateful for his insights on the matter.

27. See James C. McCann, *People of the Plow: An Agricultural History of Ethiopia* (Madison: University of Wisconsin Press, 1995), 42–44, for issues of crop choice and farm management.

28. See debate on rural class in Donald Crummey, *Land and Society in the Christian Kingdom of Ethiopia* (Urbana and Addis Ababa: University of Illinois Press and Addis Ababa University Press, 2001), 127–129, in contrast to Donald Donham, "Old Abyssinia and the New Ethiopian Empire: Themes in Social History," in *The Southern Marches of Imperial Ethiopia: Essays in History and Social Anthropology,* ed. Donald Donham and Wendy James (Cambridge: Cambridge University Press, 1986), 3–48.

29. For an insightful comparison of northern European and highland Ethiopian family and property systems, see Allan Hoben, "Family, Land, and Class in Northwest Europe and Northern Highland Ethiopia," *Proceedings of the First United States Conference on Ethiopian Studies*, 1973 (East Lansing: Michigan State University Press, 1975), 157–170.

30. There is no record of the first arrival of maize, but the Portuguese Jesuit F. Paes often receives credit. The Jesuits established an experimental farm at Fremona in Tigray to sustain their small community. See McCann, *People of the Plow*, 52.

31. Henry Salt, quoted in George Annesley Valentia, *Voyages and Travels to India, Ceylon, and the Red Sea, Abyssinia, and Egypt in the Years 1802, 1803, 1804, 1805, and 1806*, 3 vols. (London: William Miller, 1809), 3:5.

32. Antonio Cecchi, *Da Zeila alle frontiere del Caffa*, 2 vols. (Rome: Ermanno Loescher, 1886), 2:278.

33. See McCann, *People of the Plow*, 147–190, for description of the agriculture of forest ecology.

34. S. Twumasi-Afriye, a CIMMYT maize specialist, showed me in his office ten local "farmer varieties" of maize that ranged from purple to red to yellow or were variegated in color, and nine of the ten were flint types. The varieties were the result of one season's collection in a southern highland area. Interview with S. Twumasi-Afriye, 17 June 2002.

35. Raffaele Ciferri and Enrico Bartolozzi, "La produzione cerealicola dell'Africa Orientale italiana nel 1938" [Cereal Production in Italian East Africa in 1938], *L'Agricoltura Coloniale* 19, nos. 11–12 (1940): 6–7.

36. For details on the Italian demographic plans for Italian East Africa, see Haile Mariam Lerebo, *The Building of an Empire: Italian Land Policy and Practice in Ethiopia, 1935–41* (Oxford: Oxford University Press, 1994), 82–176. See also McCann, *People of the Plow*, 210–213.

37. For early U.S. aid programs, see volumes 9 and 10 of the *Ethiopian Observer*, and McCann, *People of the Plow*, 239–261.

38. See Kassahun Seyoum, Hailu Tafesse, and Steven Franzel, *The Profitability of Coffee and Maize among Smallholders: A Case Study of Limu Awraja, Ilubabor Region*, Research Report no. 10 (Addis Ababa: Institute of Agricultural Research, 1990), 10–22; McCann, *People of the Plow*, 184–186.

39. Ian Watt, "Regional Patterns of Cereal Production and Consumption," in Zein Ahmed Zein and Helmut Kloos, eds., *The Ecology of Health and Disease in Ethiopia* (Addis Ababa: Addis Ababa University Press, 1988), 121.

40. The figures went from 1.55 metric tons per hectare in 1950

(1,924,3000 metric tons total production) to 3.21 metric tons per hectare (3,815,600 metric tons total production) in 1960. L. Fenaroli, *Mais, 1946–1967* (Bergamo, Italy: Istituto per la Cerealicoltura, 1968), 17–19; 55–57; personal communication, June 1999, from Alberto Vederio and Marco Bertolini of the Istituto per la Cerealicoltura, Bergamo.

5. How Africa's Maize Turned White

1. Even expert agronomists cannot distinguish maize in the early stages of growth from sorghum.

2. Donald Moodie, *The Record, Part 1, 1649–1720* (Amsterdam: Balkema, 1960), 137, quoted in M. D. W. Jeffreys, "Who Introduced Maize into Southern Africa?" *South African Journal of Science* 63 supplement (January 1967), 27.

3. The eighteenth-century references are from Franz von Winkelmann and H. Van Reenen, *Land, Its Ownership and Occupation in South Africa* (Cape Town: Juta, 1962), as quoted in Jeffreys, "Who Introduced Maize?" 28. He quotes John Ayliff, *The Journal of John Ayliff,* Peter Hinchliff, ed. (Cape Town: Balkema, 1971), from C. B. Maclean, *A Compendium of Kafir Laws and Customs* (Mt. Choke, South Africa, 1858), 154.

4. For these early accounts, see Joseph Burtt-Davy, *Maize: Its History, Cultivation, Handling, and Uses. With Special Reference to South Africa* (London: Longman, 1914), 12–13.

5. J. A. Howard, "The Economic Impact of Improved Maize Varieties in Zambia," Ph.D. diss., Michigan State University, 1994, quoted in Melinda Smale and Thomas Jayne, *Maize in Eastern and Southern Africa: "Seeds" of Success in Retrospect,* EPTD Discussion Paper no. 97 (Washington, D.C.: Environmental and Technical Division, International Food Policy and Research Institute, January 2003, 9; Marvin Miracle, *Maize in Tropical Africa* (Madison: University of Wisconsin Press, 1966), 156.

6. David Gough, "A Reconsideration of Some Southern Bantu Maize Terms," *Essays in Bantu Language Studies,* Working Paper no. 7 (Rhodes University, South Africa), 3–5.

7. See Chapter 2. Battel is quoted in Jeffreys, "Who Introduced Maize?" 28.

8. Megan Vaughan, *The Story of an African Famine: Gender and Famine in Twentieth-Century Malawi* (Cambridge: Cambridge University Press, 1987), 52–53. Vaughan suggests that maize had been present in the area "before the eighteenth century." For a wider view of agricultural change in central Africa, see also Megan Vaughan and Henrietta Moore, *Cutting*

Down Trees: Gender, Nutrition, and Agricultural Change in Northern Rhodesia (Portsmouth, N.H.: Heinemann, 1994).

9. Melinda Smale, " 'Maize Is Life': Malawi's Delayed Green Revolution." *World Development* 23, no. 5 (1995): 819–831. Burtt-Davy reports collection of a similar flint "Kaffir" maize in southern Somalia around 1900 from an area settled by freed slaves originally from the Malawi area. Burtt-Davy, *Maize,* 314. For the dynamics of farm preference, see also Pauline Peters, "The Limits of Knowledge: Securing Rural Livelihoods in a Situation of Resource Scarcity," in C. B. Barrett, F. Place, and A. A. Aboud, eds., *Natural Resource Management in African Agriculture: Understanding and Improving Current Practices* (Cambridge, Mass.: CABI Publishing, 2002). Pauline Peters was the first to note Malawians' expression "Maize is our life."

10. William Allen, *The African Husbandman* (London: Oliver and Boyd, 1965), 68–69.

11. Audrey Richards, *Land, Labour, and Diet in Northern Rhodesia* (London: International Institute of African Language, 1939), 52.

12. Interview with Kenneth Kaunda, Boston, 8 May 2003.

13. Maize must have arrived on the South African highveld along with Indian Ocean trade contacts in the late eighteenth century. Women, its primary cultivators, must have passed along the Brazilian flint and floury types along with patterns of marriage and the household segmentation that pushed populations north and west in that period.

14. E. Casalis, *Mes souvenirs* (London: Religious Tract Society, 1882), quoted in R. C. Germond, *Chronicles of Basutoland* (Morija, Lesotho: Morija Sesuto Book Depot, 1967), 23.

15. Thomas Arbousett, *Missionary Excursion into the Blue Mountains,* David Ambrose and Albert Brutsch, eds. and trans. (Morija, Lesotho: Morija Sesuto Book Depot, 1991), 75. Here "Turkish corn" probably refers to sorghum; Indian corn is maize, and "sweet reeds" may refer to sorghum cane eaten as a snack.

16. For description of settlement and conflict see Colin Murray, *Black Mountain: Land, Class, and Power in the Eastern Orange Free State, 1880s to 1980s* (Washington, D.C.: Smithsonian Institution Press, 1992). See also Germond, *Chronicles,* 152, for a description of early Boer settlement.

17. A. F. Robertson, *The Dynamics of Productive Relationships: African Share Contracts in Comparative Perspective* (Cambridge: Cambridge University Press, 1987), 137; Colin Murray, *Families Divided: The Impact of Labour in Lesotho History* (New York: Cambridge University Press, 1981).

18. Burtt-Davy, *Maize,* 286–289. Cross-breeding later changed Hickory

King into a more productive ten- or twelve-row variety. Burtt-Davy states that Hickory King arrived on the highveld from Natal before 1898, but he offers no specific date.

19. William Beinart and Peter Coates, *Environment of History: The Taming of Nature in the USA and South Africa* (London: Routledge, 1995), 58–69. For descriptions of this dryland agronomy, see Sarah Phillips, "Lessons from the Dust Bowl: Dryland Agriculture and Soil Erosion in the United States and South Africa, 1900–1950," *Environmental History* 4 (April 1998): 245–266.

20. The spread of ox plow technology and its implications for food production and livestock management began in the 1830s in Basutoland but progressed northward early in the twentieth century. For its effects on the Tonga plateau in Northern Rhodesia (Zambia), see Kenneth Vickery, *Black and White in Southern Zambia: The Tonga Plateau Economy and British Imperialism, 1890–1939* (Westport, Conn.: Greenwood Press, 1986), 159.

21. Timothy Keegan, *Rural Transformations in Industrializing South Africa: The Southern Highveld to 1914* (Braamfontein, South Africa: Raven Press, 1986), 166–195. For a graphic narrative account of the impact on one family, see Charles Van Onselen, *The Seed Is Mine: The Life of Kas Maine, A South African Share Cropper, 1894–1985* (New York: Hill and Wang, 1996), 29–48. In the 1920s and 1930s South Africa exported almost a third of its maize crop. See Henry Bernstein, "The Boys from Bothaville, or the Rise and Fall of King Maize," paper presented to the Colloquium on Agrarian Studies, Yale University, February 1999; Robertson, *The Dynamics of Productive Relationships,* 134.

22. While men often had mine contracts allowing them to return home during the agricultural season, in practice male labor was usually missing at the key workdays for plow agriculture. John Gay, Debby Gill, and David Hall, eds., *Lesotho's Long Journey: Hard Choices at the Crossroads* (Maseru, Lesotho: Sechaba Associates, 1995), 108, presents survey data on the loss of male labor.

23. Keegan, *Rural Transformations,* 124–130. Keegan points out that mechanization in wheat production long predated that in maize.

24. Burtt-Davy, *Maize,* 322–323. Also see Beinart and Coates, *Environment and History,* 58.

25. Murray, *Families Divided,* 18.

26. For discussion of maize pricing, see Bernstein, "The Boys from Bothaville," 7. Most aggregate data on maize from South Africa reflect the situation on white commercial farms only; black smallholders are not included in national surveys.

27. South African Department of Agriculture Report, 1937, cited in Keegan, *Rural Transformation,* 119–120.

28. Yellow maize in South Africa's industrialized economy is intended for livestock or for export for industrial purposes. For statistics on white maize, see CIMMYT and FAO, *White Maize: A Traditional Food Grain in Developing Countries* (Rome: CIMMYT and FAO, 1997), 3. A new yellow version of hybrid SR-52 (see Chapter 7) is now produced for the snack maize market and as baby corn for export.

29. In a Paduan *trattoria* I asked why a restaurant in the Veneto would serve yellow polenta with a fish dish. The waiter's response was that tourists from Rome and southern Italy expected yellow.

30. Early maize in Egypt and Ethiopia was red or yellow, in Zululand blue, and in West Africa orange, red, yellow, and blue.

31. The CIMMYT gene bank has only recently shown an interest in African maize, after having argued that for a New World crop, maize varieties from Central America, Latin America, and the Caribbean are more valuable. Interview with Suketoshi Taba, Mexico City, March 1999.

32. H. Weinmann, *Agricultural Research and Development in Southern Rhodesia, 1890–1923,* Department of Agriculture Occasional Paper no. 4 (Salisbury: Rhodesia Department of Agriculture, 1972), 19–20; Smale and Jayne, *Maize in Eastern and Southern Africa,* 10–11.

33. R. C. Smith, *The Maize Story and the Farmers' Coop* (Salisbury, Rhodesia: Farmers' Seed Co-op, 1979), 150–151. Smale and Jayne's insights on this issue are especially valuable. Smale and Jayne, *Maize in Eastern and Southern Africa,* 10–11. The farmers' fear of maize's characteristic cross-pollination is reminiscent of the fears over GM types in current debates.

34. Burtt-Davy, *Maize,* 502–504. "Round whites" refers to the older flint types.

35. Only Malawi retained a preference for flint types, probably because its urban market demand was low and the community of white farmers was quite small. See chapters 4 and 5 and Smale and Jayne, *Maize in Eastern and Southern Africa,* passim.

36. See CIMMYT and FAO, *White Maize,* 7–8. Zimbabweans and Ethiopians, however, still prefer yellow maize for roasting, if they can get it.

37. Personal communication with Douglas Tanner, CIMMYT, Ethiopia, 13 May 2003; personal communication with Andrew Natsios, director of USAID, 22 April 2003.

38. Beinart and Coates, *Environment and History,* 51–71.

39. See www.cimmyt.mx/outreach/partnerssafrica.htm. The International Center for Maize and Wheat Improvement (CIMMYT) also reports that it

(rather than South Africa commercial maize breeders) has taken the lead in developing new cultivars resistant to drought and insects, as well as tolerant of low nitrogen levels.

40. Lesotho National Archives, cited in Tumelo Tsikoane, "A History of Public Health in Lesotho, Southern Africa, 1900–1980," Ph.D. diss., Department of History, Boston University, 1996.

6. African Maize, American Rust

1. For details on East African maize types and local variations, see J. D. Acland, *East African Crops: An Introduction to the Production of Field and Plantation Crops in Kenya, Tanzania, and Uganda* (Nairobi: FAO and Longman, 1971).

2. See James C. McCann, "Maize and Grace: Corn and Africa's Changing Landscapes, 1500–1999," *Comparative Studies in Society and History* 43, no. 2 (April 2001): 246–272, and map, fig. 1.1 in chap. 1.

3. Dr. G. A. C. Herklots (secretary of state for colonial agricultural research) to D. Rhind (secretary for agriculture and forestry research, West African Inter-Territorial Secretariat, Accra, Public Record Office, London Colonial Office Archives (hereafter CO), 24 July 1951, CO 927/189/7.

4. For information on the Spanish flu in Africa and the role of disease in colonial rule, see David Patterson, "The Influenza Epidemic of 1918–19 in the Gold Coast," *Transactions of the Historical Society of Ghana* 16 (1977).

5. "Rust (fungus)," *Microsoft Encarta Encyclopedia 99*.

6. Rusts are fungi, which propagate via spores that travel on the wind or on living plant tissue, meanwhile alternating between uredeospore and teleutospore stages of virulence and dormancy, respectively. For the information I am grateful to my colleague Gillian Cooper-Driver, professor emeritus, Boston University and the National University of Lesotho.

7. C. L. M. Van Eijnatten, *Towards the Improvement of Maize in Nigeria* (Wageningen, Netherlands: University of Wageningen, 1965), 70; D. Rhind, J. M. Waterson, and F. C. Deighton, "Occurrence of *Puccinia polysora Underw.* in West Africa," *Nature* 1969 (1952): 631; for Gold Coast losses, see D. Rhind, Report to Secretary for Agricultural and Forestry Research, West-African Inter-Territorial Secretariat, Accra, 24 October 1951, CO 927/189/7. Estimates of losses seem to have been guesswork, since colonial agricultural officers had no mechanisms for crop production beyond the losses to rust on test plots.

8. D. Rhind, secretary for agricultural and forestry research, "Memo-

randum on Maize Rust Disease in West Africa," 18 June 1951, transmitted by chief secretary's deputy, West African Inter-Territorial Secretariat to governor, Nigeria; OAG, Gold Coast; governor, Sierra Leone; OAG, Gambia, CO 927/189/7.

9. Ibid.

10. Ibid.

11. The first meeting of the Maize Rust Research Unit took place on 4 March 1953. G. A. C. Herklots, "Rust Disease of Maize in West Africa," 18 July 1951. Geoffrey Herklots was the secretary for the Committee for Colonial Agriculture, Animal Health, and Forestry Research, founded in June 1945. Significantly, the Moor Plantation Research Station eventually evolved into the International Institute for Tropical Agriculture (IITA), part of the current Consultative Group for International Agricultural Research (CGIAR) system. For actions of the Maize Rust Research Unit, see CO 927/277.

12. Rhind to Wiltshire, 18 June 1951, CO 927/189/7.

13. Webster to Herklot (Colonial Office), 27 July 1951, CO 927/189/7. The 1948 Economic Cooperation Act established the ECA as a branch of the State Department that would administer the European Recovery Program (Marshall Plan). The ECA eventually evolved into the United States Agency for International Development (USAID) in 1961. My thanks to Sarah Phillips for this information.

14. Jenkins was in charge of all government maize-breeding programs in the United States.

15. Herklots to Rhind, 24 July 1951, and Herklots Minute of 29 October 1951, CO 927/189.

16. Rhind, Deighton, and Waterston, "Occurrence of *Puccinia polysora Underw.* in West Africa," 631.

17. Beginning in 1943, the Rockefeller Foundation had spearheaded efforts in international crop research in Mexico that eventually resulted in the mid-1960s in the formation of the worldwide organization CGIAR. See Bruce Jennings, *Foundations of International Agricultural Research: Science and Politics in Mexican Agriculture* (Boulder, Colo.: Westview, 1988), 9–44.

18. W. R. Stanton and R. H. Cammack, "Resistance to the Maize Rust *Puccinia polysora Underw.,*" *Nature* 172 (1953): 505–506.

19. Waterston to Rhind, 24 September 1953, CO 927/277.

20. Committee for Colonial Agricultural, Animal Health, and Forestry Research, 3 November 1953, CO 927/276.

21. Stanton and Cammack, "Resistance to the Maize Rust," 505–506.

22. Norman Borlaug, "A Case of Stable Balanced Biotic Relationship between Maize and Its Two Rust Parasites," *Proceedings of the Conference on*

the Biology of Rust Resistance in Forest Trees (Washington, D.C.: Forest Service, U.S. Department of Agriculture, 1969), 618–619.

23. Ibid.

24. Van Eijnatten, *Towards the Improvement of Maize,* 5.

25. Borlaug, "A Case," 619. Here he ignores evidence from African colonial research efforts on test plots that no African maize types showed any resistance.

26. I am grateful to Brian Spooner, head mycologist at the Royal Botanic Gardens at Kew for this example. James Webb pointed out the case of the "brown bug" from his work on Sri Lanka. See James L. A. Webb, Jr., *Tropical Pioneers: Human Agency and Ecological Change in the Highlands of Sri Lanka, 1800–1900* (Athens: Ohio University Press, 2002), 86–87.

27. Colin Trapnell and J. N. Clothier, *The Soils, Vegetation, and Agricultural Systems of Northwestern Rhodesia: Report of the Ecological Survey* (Lusaka, Northern Rhodesia: Government Printer, 1957); Colin Trapnell, *Ecological Survey of Zambia: The Traverse Records of C. G. Trapnell, 1932–1943,* 3 vols., Paul Smith, ed. (London: Royal Botanic Gardens at Kew, 2001); John Ford, *The Role of the Trypanosomiases in African Ecology: A Study of the Tsetse Fly Problem* (Oxford: Clarendon Press, 1971); William Allen, *The African Husbandman* (Edinburgh: Oliver and Boyd, 1965), passim.

28. My thanks to Steven Feierman for suggesting this line of argument. For information on these programs, see Feierman, *Peasant Intellectuals: Anthropology and History in Tanzania* (Madison: University of Wisconsin Press, 1990); David Anderson, "Depression, Dustbowl, Demography, and Drought: The Colonial State and Soil Conservation in East Africa during the 1930s," *African Affairs* 83 (1984): 321–343; Kirk Hoppe, "Lords of the Flies: British Sleeping Sickness Policies as Environmental Engineering in the Lake Victoria Region, 1900–1950," African Studies Center Working Paper no. 203, African Studies Center, Boston University, 1995.

7. Breeding SR-52: The Politics of Science and Race in Southern Africa

1. The name was, paradoxically, an anachronism, since in 1953 the self-governing colony of Southern Rhodesia had become part of the commonwealth territory called the Federation of Rhodesia and Nyasaland.

2. I am tempted to name Secretariat, the legendary Triple Crown winner, but that horse was less effective than SR-52 in siring the next generation.

3. Interview with Mike Caulfield, 17 July 2003. Caulfield is a maize

breeder who joined the Salisbury Agricultural Research Station in 1960 and later worked for SeedCo, Zimbabwe's private seed research corporation. Kingstone Mashingaidze, "Maize Research and Development," in Mandivamba Rukuni and Carl Eicher, eds., *Zimbabwe's Agricultural Revolution* (Harare: University of Zimbabwe, 1997), 212–213.

4. For an overview of the demography of early settlement, see, for example, David Beach, *The Shona and Their Neighbors* (Oxford: Blackwell, 1994), 43–49, 177–280; Ian Phimister, *An Economic and Social History of Zimbabwe, 1890–1948: Capital Accumulation and Class Struggle* (London: Longwood, 1988), 57–80.

5. Like the contemporaneous Aswan High Dam, the Kariba Dam received most of its funding from the World Bank and other international sources of capital. An excellent narrative history of this period is Eugenia Herbert, *Twilight on the Zambezi: Late Colonialism in Central Africa* (New York: Palgrave, 2002), 90–91.

6. Louis Gann, *Central Africa: The Former British States* (Englewood Cliffs, N.J.: Prentice Hall, 1971), 19–20.

7. Personal communication with Tony O'Neill, who served in the Labor Department of the Rhodesian Ministry of Agriculture from 1962 until his retirement; interview with Mike Caulfield, Harare, 17 July 2002.

8. R. C. Smith says that a visiting American agriculturalist named Odlam recommended importation of improved varieties from South Africa, while the Department of Agriculture brought varieties directly from the United States. R. C. Smith, *The Maize Story and the Farmer's Coop* (Salisbury, Rhodesia: n.p., 1979), 144. See also *Rhodesia Journal of Agriculture* 11 (1913).

9. These were resistant to the older leaf blight, but not to the gray leaf spot that ravaged maize crops in the 1990s.

10. Mashingaidze, "Maize Research and Development," 210. The Botanical Experimental Station, later the Salisbury Agricultural Research Station, eventually became the Harare Research Station.

11. Interview with Mike Caulfield, Harare, 17 July 2003.

12. For an insightful analysis of the colonial government's views on African agriculture, see William Munro, "To Civilize Both the Land and the People: Governmentality and the Environment in Late-Colonial Zimbabwe," paper presented to the American Society for Environmental History, Providence, R.I., March 2003. One might relate this emphasis in Southern Rhodesia to American maize breeding, which also focused on the needs of commercial farmers, rather than those of smallholders and sharecroppers less attuned to the market.

13. For these patterns in the various colonies, see Melinda Smale and Thomas Jayne, *Maize in Eastern and Southern Africa: "Seeds" of Success in Retrospect,* EPTD Discussion Paper no. 97 (Washington, D.C.: Environmental and Technical Division, International Food Policy and Research Institute, January 2003), 12–14; Kenneth Vickery, *Black and White in Southern Zambia: The Tonga Plateau Economy and British Imperialism, 1890–1939* (Westport, Conn.: Greenwood Press, 1986), 195–200; and, for Kenya, Gavin Kitching, *Class and Economic Change in Kenya: The Making of an African Petite-Bourgeoisie* (New Haven: Yale University Press, 1980), 57–100, 315–374. Paul Moseley makes the same point in *The Settler Economies: Studies in the Economic History of Kenya and Southern Rhodesia, 1900–1963* (Cambridge: Cambridge University Press, 1983).

14. See Smith, *The Maize Story,* 158–165. Smith's account of the board, written just before Zimbabwe gained black majority rule, is an apologia for the white farmers' position.

15. Smale and Jayne, *Maize in Eastern and Southern Africa,* 13.

16. Ibid., 13–14.

17. An excellent, and perhaps underappreciated, account of Zambian agriculture in this period is Vickery, *Black and White in Southern Zambia,* 195–200.

18. See Penny Grant, "Obituary: Alan George Hay Rattray," *The Farmer,* 28 May 1998.

19. It is difficult to ascertain how breeders responded to the implications for black smallholders in 1949; in the case of 1969, as we shall see, they were surprised by black farmers' response to hybrids but reacted by trying to make seeds more available to the African smallholders.

20. See Munro, "To Civilize," 12.

21. Malawi was very much the junior partner in the federation, and its small European population had far less influence than did Southern Rhodesia and Northern Rhodesia on federation politics and resource allocation. The Nyasaland maize-breeding research efforts were small and were damaged by the 1959 resignation of their expatriate staff.

22. Interview with Mike Caulfield, Harare, 17 July 2003. Caulfield was reporting on events that had happened the year before his arrival but had become part of SR-52's local narrative.

23. Smith, *The Maize Story,* 65.

24. This idea was suggested to me by CIMMYT agronomist Stephan Waddington, who has worked with farmers in Zimbabwe since 1984.

25. Rex J. Tattersfield and Ephraim K. Havazvidi, "The Development of the Seed Industry," in Rukuni and Eicher, *Zimbabwe's Agricultural Revolu-*

tion, 117. The authors cite A. G. H. Rattray, "Maize in Rhodesia, the Cultivation of the Crop," *Rhodesia Agricultural Journal* 10 (1970): 4–15.

26. For two overviews of the federation's collapse, see Herbert, *Twilight on the Zambezi,* and Harry Franklin, *Unholy Wedlock: The Failure of the Central African Federation* (London: Allen and Unwin, 1963).

27. Patrick H. Tawonezvi, "Agricultural Research Policy," in Rukuni and Eicher, *Zimbabwe's Agricultural Revolution,* 92.

28. See Mashingaidze, "Maize Research and Development," 210–211; Smale and Jayne, *Maize in Eastern and Southern Africa,* 19–20; William Masters, *Government and Agriculture in Zimbabwe* (Westport, Conn.: Greenwood, 1994), 56. A more recent, early-maturing single-cross (ZS225) failed in zones of marginal rainfall because of its particular sensitivity to low moisture levels at the tasseling stage.

29. Interview with Mike Caulfield, 17 July 2003. Figure 27 is a copy of his original chart showing the market anomaly. Bernard Kupfuma refers to the post-1980 period as Africa's second maize revolution. See Carl Eicher and Bernard Kupfuma, "Zimbabwe's Emerging Maize Revolution," in Derek Byerlee and Carl Eicher, eds., *Africa's Emerging Maize Revolution* (Boulder, Colo.: Lynne Rienner, 1997), 25–27. Caulfield reports that once they realized what was happening, they began producing smaller packs of seed, to suit the needs of smallholders.

30. Mashingaidze, "Maize Research," 212, 215.

31. Byerlee and Eicher, "Africa's Emerging Maize Revolution," 25–26.

32. For an analysis of the effects of drought in the mid-1980s, see Michael Bratton, "Drought, Food, and Social Organization," in Michael Glantz, ed., *Drought and Hunger in Africa: Denying Famine a Future* (Cambridge: Cambridge University Press, 1987), 245–268.

33. This has long been the thesis of Carl Eicher's agricultural economist team at Michigan State University and the University of Zimbabwe; see Byerlee and Eicher, *Africa's Emerging Maize Revolution.* A critique later emerged within the same group, which has examined the data from a social and political perspective. See "Abstract," in Smale and Jayne, *Maize in Eastern and Southern Africa,* iii. For maize sales from 1979 to 1990, see Appendix Table 7.1 in this book.

34. Mashingaidze, "Maize Research," 194, 215.

35. The first young men from Tonga to bring a plow from Southern Rhodesia did so in 1902; by the 1920s the Oliver plow was in widespread use. That it was predominantly an instrument used by men meant that the expanding maize fields of the 1920s and 1930s were male turf. Vickery, *Black and White,* 159–169.

36. The best secondary source on this transformation is Vickery, *Black and White;* one of the most insightful primary sources about the rural environment and the rural economy is C. G. Trapnell and J. N. Clothier, *The Soils and Vegetation of Northern Rhodesia (Report of the 1932 Ecological Survey)* (Lusaka, Zambia: Government Printer, 1937), which describes in great detail small-scale agriculture in transition throughout Northern Rhodesia in the 1930s.

37. Vickery, *Black and White,* 169.

38. Interview with Kenneth Kaunda, 8 May 2003.

39. Vickery, *Black and White,* 219.

40. Interview with Kenneth Kaunda, 8 May 2003. Although Kaunda readily avowed his suspicions of GM maize, which could be linked to external control, he viewed SR-52 as a great boon to the nation.

41. Ibid.; Herbert, *Twilight on the Zambezi,* 100. Here Kaunda differed substantially from the more urban orientation of his contemporaries among the nationalist leaders.

42. Julie A. Howard and Catherine Mungoma, "Zambia's Stop-and-Go Maize Revolution," in Byerlee and Eicher, *Africa's Emerging Maize Revolution,* 48–49.

43. See Howard and Mungoma, "Zambia's Stop-and-Go Maize Revolution," 49–50. This contamination was detected only after 1977, though its effects were evident throughout the 1970s.

44. Interview with Kenneth Kaunda, 8 May 2003.

45. D. Ristanovic and P. Gibson, "Development and Evaluation of Maize Hybrids in Zambia," in Brhane Gelaw, ed., *To Feed Ourselves: A Proceedings of the First Eastern, Central, and Southern African Regional Workshop,* Lusaka, Zambia, March 1985 (Mexico City: CIMMYT, 1985).

46. Howard and Mungoma, "Zambia's Stop-and-Go Maize Revolution," 50.

47. Christopher Dowswell, R. L. Paliswal, and Ronald P. Cantrell, *Maize in the Third World* (Boulder, Colo.: Westview, 1996), 8. For an in-depth study of Malawi, especially the Shire Valley, see Pauline Peters, "The Limits of Knowledge: Securing Rural Livelihoods in a Situation of Resource Scarcity," in C. B. Barrett, F. Place, and A. A. Aboud, eds., *Natural Resource Management in African Agriculture: Understanding and Improving Current Practices* (Cambridge, Mass.: CABI Publishing, 2002), 12.

48. I am grateful to Dr. Peters for sharing the draft chapter "The Limits of Knowledge," based on her research, with me. Peters's work on the Shire highlands complements work on the sorghum-based Shire Valley by the historian Elias Mandala. See Elias Mandala, *Work and Control in a Peasant*

Economy: A History of the Lower Tchiri Valley in Malawi, 1859–1960 (Madison: University of Wisconsin Press, 1990), 161–190. The work on the Shire Valley complements work done at the national scale over several years by agricultural economist Melinda Smale, who has sought to understand in depth the background for Malawi's distinctive maize economy.

49. Peters, "Limits of Knowledge," 13.

50. Smale and Peters agree that nationally in 1986 only 5 percent of farms were sowing hybrid maize, compared with almost 100 percent in Zimbabwe and South Africa, and about 70 percent in Zambia.

51. Peters, "The Limits of Knowledge," 21.

52. Interview with maize breeder Xavier Mhike, Harare Research Station, Harare, 14 July 2003.

53. Paul Richards, "Ecological Change and the Politics of African Land Use," *African Studies Review* 26, no. 2 (1983): 27, quoted in Peters, "The Limits of Knowledge"; see also John Illife, *Africa: History of a Continent* (Cambridge: Cambridge University Press), 27.

54. Peters, "The Limits of Knowledge," 26–27.

55. The best survey of Kenya's rural economic history is Gavin Kitching, *Class and Economic Change in Kenya*, 25–158.

56. Ibid., 25, 29.

57. Smale and Jayne, *Maize in Eastern and Southern Africa*, 9.

58. See John Gerhart, *The Diffusion of Hybrid Maize in Western Kenya* (Mexico City: CIMMYT, 1975), 2.

59. Gerhart, *Diffusion*, 4–5. Also see Smale and Jayne, *Maize in Eastern and Southern Africa*, 16–17.

60. For exact figures on maize production in Kenya, Malawi, Zambia, and Zimbabwe from 1990 to 2000, see Appendix Table 7.2; Smale and Jayne, *Maize in Eastern and Southern Africa*, 44.

8. Maize and Malaria

1. Bekele Abebe operated a private pharmacy and was the Burie district's only health-care provider from 1968 until 2000. He reports that malaria cases in the district were extremely rare and consisted only of patients from the lowlands where malaria is endemic. Interview with Bekele Abebe, Addis Ababa, 24 May 2003.

2. Interviews with Semahagn Abate, secretary of the Antimalaria Association, Burie chapter, May 2002.

3. My partner in this research has been Asnakew Kebede of the Ethiopian Ministry of Health and the World Health Organization, who was in-

strumental in assembling the data from local health clinics and working with me to formulate hypotheses on the interaction of maize and malaria. Our joint work appears in our conference paper "Maize and Malaria: A New Connection. History, Disease, and Agro-Ecology in the Deadly Gojjam Epidemic of 1998," presented to the annual conference of the American Society for Environmental History, Providence, R.I., March 2003.

4. Three published studies of laboratory work on mosquito links to maize pollen now exist. See Yemane Ye-ebiyo, Richard Pollack, and Andrew Spielman, "Enhanced Development in Nature of Larval *Anopheles arabiensis* Mosquitoes Feeding on Maize Pollen," *American Journal of Tropical Medicine and Hygiene* 63, nos. 1–2 (2000): 91–92; Yemane Ye-ebiyo, Richard Pollack, Anthony Kiszewski, and Andrew Spielman, "Enhancement of Development of Larval *Anopheles arabiensis* by Proximity to Flowering Maize *(Zea mays)* in Turbid Water and When Crowded," *American Journal of Tropical Medicine and Hygiene* 68, no. 6 (2003): 748–752; Yemane Ye-ebiyo, Richard Pollack, Anthony Kiszewski, and Andrew Spielman, "A Component of Maize Pollen That Stimulates Larval Mosquitoes *(Diptera culicidae)* to Feed and Increases Toxicity of Microbial Larvicides," *Journal of Medical Entomology* 40, no. 6 (November 2003): 860–864.

5. B. Ameneshewa and M. W. Service, "The Relationship between Female Body Size and Survival Rate of the Malaria Vector *Anopheles arabiensis* in Ethiopia," *Medical Veterinary Entomology* 10 (1996): 170–172.

6. Actually, Andrew Spielman first got in touch with me, having heard of my research project on maize in Africa. I then proposed the research site at Burie, and he arranged for me to meet my research partner Asnakew Kebede.

7. In May 2003 I returned to the Burie district to gather further case evidence of malaria and to assess the impact of a rare failure of the main rains in the previous season. This drought was the first in the West Gojjam region for many years.

8. The Bir Sheleko farm is in a flat-bottomed valley ideally suited to its large-scale mechanized maize production operation. Its raw figures for temperature and rainfall would differ somewhat from those of the adjacent Burie highlands, but the annual patterns, we argue, would be consistent.

9. The best descriptions of these patchwork patterns of settlement in northeast Ethiopia are still to be found in Allan Hoben, *Land Tenure among the Amhara of Ethiopia: The Dynamics of Cognatic Descent* (Berkeley: University of California Press, 1974), 32–65, and Frederick Simoons, *North-*

west Ethiopia: Peoples and Economy (Madison: University of Wisconsin Press, 1960), 57–84.

10. Yeshanew Ashagrie, Matts Olsen, and Tekalign Mamo, "Contribution of *Croton macrostachys* to Soil Fertility in Maize-Based Subsistence Agriculture of Bure [Burie] Area, North-Western Ethiopia," *Maize Production Technology for the Future: Challenges and Opportunities. Proceedings of the Sixth Eastern and Southern Africa Regional Maize Conference* (Mexico City: CIMMYT, 1999), 232–234.

11. For a quick and accurate reference on mosquito species and malaria, see Andrew Spielman, *Mosquito: A Natural History of Our Most Persistent and Deadly Foe* (New York: Hyperion, 2001), 17–29.

12. An example of external introduction was the 1889 outbreak of rinderpest, brought from India by the Italian importation of zebu cattle. See James C. McCann, *From Poverty to Famine in Northeast Ethiopia: A Rural History* (Philadelphia: University of Pennsylvania Press, 1985).

13. The absence of malaria and any mosquito vector corresponded to my personal observation after two years' residence in the area in 1973–1975, and also to that of local medical professional Bekele Abebe, who served the area from 1968 to 2000 (interview with Bekele Abebe, 24 May 2002). Also see Asnakew Kebede, "Overview of the History of Malaria Epidemics in Ethiopia," paper presented at the Workshop on Capacity Building on Malaria Control in Ethiopia, Addis Ababa, March 2002.

14. I am grateful to Abdusamad Hajj Ahmed for sharing his family's experience with malaria with me.

15. Interview with Getneh Atres, age fifty-three, Imborbor, Gulim, 24 May 2002.

16. See Yemane, Pollack, and Spielman, "Enhanced Development," 92.

17. See McCann, *People of the Plow,* chap. 8. The late Ian Watt, geographer at Addis Ababa University, was the first person to call my attention to growing role of maize in Ethiopian agriculture.

18. See James Feeley and Ian Scoones, "Knowledge, Power and Politics: The Environmental Policy-Making Process in Ethiopia," *Journal of Modern African Studies* 38, no. 1 (2000): 97–101.

19. For a description of the problems associated with the SG2000 package elsewhere in Ethiopia, see Teferi Abate, "Government Intervention and Socioeconomic Change in a Northeast Ethiopian Community: An Anthropological Study," Ph.D. diss. Department of Anthropology, Boston University, 2000.

20. This figure is based on informal field surveys conducted in May 2002 in Gulim, Zeyaw Showen, and Fetam Sentom. In each of these places maize cultivation was especially concentrated around new housing sites.

21. Interview with Getneh Atres, Imborbor, Gulim, 24 May 2002.

22. Ibid.

23. See Betty Fussell, *The Story of Corn: The Myths and History, the Culture and Agriculture, the Art and Science of America's Quintessential Crop* (New York: Knopf, 1992), 3.

24. Market observations in May 2002 indicated that older, multicolored composite varieties are still grown, but in much smaller quantities than BH660. Large grain merchants accepted only the pure-white BH660 maize, while the older types were sold in smaller amounts for local consumption and for seed.

25. Tesfaye Tsegaye and Alemu Hailye, *Adoption of Improved Maize Technologies and Inorganic Fertilizer in Northwestern Ethiopia* (Addis Ababa: Institute for Agricultural Research, 2001), 38.

26. Gojjam's nongovernmental organization the Antimalaria Association unofficially tallied 222,992 cases and 7,783 deaths, but it estimated that the actual numbers were twice that (personal communication from Semahagne Abate, Burie secretary of the Antimalaria Association).

27. The Ministry of Agriculture taxonomy agreed with our own assessment of the areas we assessed directly and their involvement in maize production.

9. Maize as Metonym in Africa's New Millennium

1. P. L. Pingali and S. Pandey, "Meeting World Maize Needs: Technological Opportunities and Priorities for the Public Sector," part 1 in P. L. Pingali, ed., *CIMMYT 1999–2000 World Maize Facts and Trends. Meeting World Maize Needs: Technological Opportunities and Priorities for the Public Sector* (Mexico City: CIMMYT, 2001), 1.

2. Much of the gender-related nature of maize adoption is open to informed speculation. For more solidly reported cases, see Kenneth Vickery for Tonga and Margaret Jean Hay on the Luo: Kenneth Vickery, *Black and White in Southern Zambia: The Tonga Plateau and British Imperialism, 1890–1939* (Westport, Conn.: Greenwood Press, 1986); Margaret Jean Hay, "Economic Change in Luoland: Kowe, 1890–1945," Ph.D. diss., Department of History, University of Wisconsin, Madison, 1971.

3. Personal communication, Cosmos Magorokosho, doctoral student at Texas A&M University, and Marianne Bänziger, CIMMYT-Harare.

4. James Scott, *Seeing Like a State: How Certain Schemes to Improve the Human Condition Have Failed* (New Haven, Conn.: Yale University Press, 1998).

5. David Miller and Wally Marasas, "Ecology of Mycotoxins in Maize and Groundnuts," in ILEIA, *Ecosystem Disruption and Human Health* (Leusden, Netherlands: Centre for Information on Low External Input Sustainable Agriculture, 2002), 23.

6. Ibid., 24.

7. Ibid.

8. For a helpful and balanced discussion of the evidence from fieldwork among Mexican farmers and the genetics of GM maize, see Mauricio R. Bellon and Julien Berthaud, "Exploring the Potential Impacts of Transgenic Varieties on Maize Diversity in Traditional Mexican Agricultural Systems," paper prepared for the Colloquium Series, Program in Agrarian Studies, Yale University, November 2003.

9. Erika Meng and Javier Ekboir, "Current and Future Trends in Maize Production and Trade," in Pingali and Pandey, "Meeting World Maize Needs," 35–37.

10. Personal communication from Wilfred Mwangi, CIMMYT economist for East Africa, June 1999.

11. Joachim Voss, director general of the International Center for Tropical Agriculture (Cali, Colombia), personal communication, 19 July 2002.

12. For the project philosophy of the Sasakawa Global 2000 program in Ethiopia, see James Feeley and Ian Scoones, "Knowledge, Power and Politics," *Journal of Modern African Studies* 38, no. 1 (2000): 99–101. The Sasakawa Global 2000 package has been impressive (see Chapter 7), but it is narrowly focused on short-term gains. In 1999 I met with Benti Tolossa, Ethiopia's most experienced maize breeder. He spoke with enthusiasm about the presence of old abandoned varieties of maize in pockets of Ethiopia, but he had followed the immediate imperatives of developing quality protein maize from modern breeding lines.

13. Pingali and Pandey, "Meeting World Maize Needs," 1. For Addis Ababa cattle fattening see James C. McCann, *People of the Plow: An Agricultural History of Ethiopia* (Madison: University of Wisconsin Press, 1995), 235–238.

Select Bibliography

Acland, J. D. *East African Crops: An Introduction to the Production of Field and Plantation Crops in Kenya, Tanzania, and Uganda.* Nairobi: Food and Agriculture Organization of the United Nations (FAO) and Longman, 1971.

Amadei, Giorgio. "Dalla polenta alla plastica: Maize: from Polenta to Plastics." In Roberto Anderlini, Giorgio Amadei, and Roberto Bartolini, eds., *Il mais: On Maize.* Milan: Agrimont, 1989.

Anderlini, Roberto. "Scoperto e diffusione del mais: The History of Maize." In Roberto Anderlini, Giorgio Amadei, and Roberto Bartolini, eds., *Il mais: On Maize.* Milan: Agrimont, 1989.

Asnakew Kebede, "Overview of the History of Malaria Epidemics in Ethiopia." Paper presented at the Workshop on Capacity Building on Malaria Control in Ethiopia. Addis Ababa, March 2002.

Asnakew Kebede and James C. McCann. "Maize and Malaria: A New Connection. History, Disease, and Agro-Ecology in the Deadly Gojjam Epidemic of 1998." Paper presented to the annual conference of the American Society for Environmental History, Providence, R.I., March 2003.

Bänziger, Marianne, and Julien de Meyer, "Collaborative Maize Variety Development for Stress-Prone Environments in Southern Africa." In D. A. Cleveland and D. Soleri, eds., *Farmers, Scientists and Plant Breeding.* Harare, Zimbabwe: CAB International, 2002.

Bartolini, Roberto. "Dal trinciato all granella: From Silage to Grain." In Roberto Anderlini, Giorgio Amadei, and Roberto Bartolini, eds., *Il mais: On Maize.* Milan: Agrimont, 1989.

———. "Del seme alla spiga: Maize Cultivation Techniques." In Roberto Anderlini, Giorgio Amadei, and Roberto Bartolini, eds., *Il mais: On Maize.* Milan: Agrimont, 1989.

Bellon, Mauricio. "The Ethnoecology of Maize Variety Management: A Case Study from Mexico." *Human Ecology* 19, no. 3 (1991): 389–418.

Bellon, Mauricio, and Julien Berthaud. "Exploring the Potential Impacts of Transgenic Varieties on Maize Diversity in Traditional Mexican Agricultural Systems." Paper prepared for the colloquium series, Program in Agrarian Studies, Yale University, November 2003.

Bellon, Mauricio, and J. Edward Taylor. " 'Folk' Soil Taxonomy and the Partial Adoption of New Seed Varieties." *Economic Development and Economic Change* 41, no. 4 (1993): 763–786.

Benti Tolessa and Joel K. Ransom, eds. *Proceedings of the First National Maize Workshop of Ethiopia.* Addis Ababa: Institute of Agricultural Research and International Maize and Wheat Improvement Center [CIMMYT], 1993.

Bernstein, Henry. "The Boys from Bothaville, or the Rise and Fall of King Maize." Paper presented to the Colloquium on Agrarian Studies, Yale University, February 1999.

Blench, Roger, Kay Williamson, and Bruce Connell. "The Diffusion of Maize in Nigeria: A Historical and Linguistic Investigation." *Sprache und Geschichte in Afrika* 15 (1994): 9–46.

Borlaug, Norman. "A Case of Stable Balanced Biotic Relationship between Maize and Its Two Rust Parasites." In *Proceedings of the Conference on Rust Resistance in Forest Trees.* Washington, D.C.: Forest Service, U.S. Department of Agriculture, 1969.

Borsatto, Evaristo. "Grano, granoturco e riso" [Wheat, Maize, and Rice]. In Giovan Battista Pellegrini, ed., *I lavori dei contadini.* Vicenza, Italy: Neri Pozza, 1997.

Brandolini, Aureliano. "Contributo allo studio delle varietà italiane di mais. I. Il granoturco 'Rostrato' (Zea mays L. cultivar 'rostrato')." Abstract from *Annali della Sperimentazione Agraria,* n.s. Rome: Ministero dell'Agricultura e delle Foreste, 1954.

Burtt-Davy, Joseph. *Maize: Its History, Cultivation, Handling, and Uses. With Special Reference to South Africa.* London: Longmans, Green, 1914.

Byerlee, Derek, and Carl Eicher. "Accelerating Maize Production: Synthesis." In Derek Byerlee and Carl Eicher, eds., *Africa's Emerging Maize Revolution.* Boulder, Colo.: Lynne Rienner, 1997.

———, eds. *Africa's Emerging Maize Revolution.* Boulder, Colo.: Lynne Rienner, 1997.

Byerlee, Derek, and Paul Heisey. "Past and Potential Impacts of Maize Research in Sub-Saharan Africa: A Critical Assessment." *Food Policy* 21, no. 3 (1996): 255–277.

Carney, Judith. *Black Rice: The African Origins of Rice Cultivation in the Americas.* Cambridge, Mass.: Harvard University Press, 2001.

Carter, George. "Maize to Africa." *Anthropological Journal of Canada* 1, no. 2 (1963): 3–8.

Centro Internacional de Mejoridad de Maíz y Trigo (CIMMYT). *A Sampling of CIMMYT Impacts, 1998: Ten Case Studies.* Mexico City: CIMMYT, 1998.

———. *Seed Conservation and Distribution: The Role of the CIMMYT Maize Germplasm Bank.* Mexico City: CIMMYT, 1986.

———. *The Quality Protein Maize Revolution: Improved Nutrition and Livelihoods for the Poor.* Mexico City: CIMMYT, n.d.

———. *World Maize Facts and Trends, 1991–1992.* Mexico City: CIMMYT, 1992.

———. *World Maize Facts and Trends, 1997–1998.* Mexico City: CIMMYT, 1999.

Centro Internacional de Mejoridad de Maíz y Trigo (CIMMYT).and Food and Agriculture Organization of the United Nations (FAO). *White Maize: A Traditional Food Grain in Developing Countries.* Rome: CIMMYT and FAO, 1997.

Ciferri, Raffaele, and Enrico Bartolozzi. "La produzione cerealicola dell'Africa Orientale italiana nel 1938" [Cereal production in Italian East Africa in 1938]. *L'Agricoltura Coloniale* (Florence) 19, nos. 11–12 (1940): 6–7.

CIMMYT. *See* Centro Internacional de Mejoridad de Maíz y Trigo (CIMMYT).

Ciriacono. Salvatore. "Introduction." In Salvatore Ciriacono, ed., *Land Drainage and Irrigation.* Aldershot, England: Ashgate, 1998.

Coppola, Gauro. *Il mais nell'economia agricola lombarda (dal secolo XVII all'Unità)* [Corn in the Agricultural Economy of Lombardy (from the Seventeenth Century to the Unification)]. Bologna: Il Mulino, 1979.

Cordova, Hugo. "Quality Protein Maize: Improved Nutrition and Livelihoods for the Poor," in Maize Program, *Maize Research Highlights, 1999–2000.* Mexico City: CIMMYT, 2001, 27–31.

Coursey, D. G. *Yams: An Account of the Nature, Origins, Cultivation, and Utilization of the Useful Members of the Dioscoreaceae.* London: Longmans, 1967.

Crosby, Alfred. *The Columbian Exchange: Biological and Cultural Consequences of 1492.* Westport, Conn.: Greenwood Press, 1972.

———. *Ecological Imperialism: The Biological Expansion of Europe, 900–1900.* Cambridge: Cambridge University Press, 1986.

Curatola, Marco. "Dioses y Hombres del Maíz: Religion, Agricultura, y Sociedad en el Antiguo Perú" [Gods and Men of Corn: Religion, Agriculture, and Society in Ancient Peru]. In Marco Curatola and Fernando Silva-Santisteban, eds., *Historia y Cultura del Perú* . . . Lima: Universidad de Lima, n.d.

Cutler, Hugh. "Races of Maize in South America." *Botanical Museum Leaflets, Harvard University* 12, no. 8 (1946): 257–291.

De Groote, Hugo. "Economic Analysis and Impact Assessment of QPM— An Overview." Paper based on a presentation for the Nippon Foundation Project meeting, Nairobi, April 2002.

De Marees, Pieter. *Description and Historical Account of the Gold Kingdom of Guinea (1602)*. Translated from the Dutch by Albert van Dantzig and Adam Jones. Oxford: Oxford University Press, 1987.

Dowswell, Christopher, R. L. Paliswal, and Ronald P. Cantrell. *Maize in the Third World*. Boulder, Colo.: Westview, 1996.

Ehret, Christopher. *An African Classical Age: Eastern and Southern Africa in World History, 1000 b.c. to a.d. 400*. Charlottesville: University Press of Virginia, 1998.

———. "Agricultural History in Central and Southern Africa, ca. 1000 b.c. to a.d. 500." *Transafrican Journal of History* 4, nos. 1–2 (1974): 1–25.

———. "Compendium of East African Culture Vocabularies." Section 1: "Nilotes, Kuliak, Central Sudanic, Southern Cushitic, and Khoisan, vol. 3." Unpublished typescript.

Eicher, Carl K. "Zimbabwe's Maize-Based Green Revolution: Preconditions for Replication." *World Development* 23, no. 5 (1995): 805–818.

Ellis, T. R. "The Food Properties of Flint and Dent Maize." *East African Agriculture* 24 (1959): 251–263.

FAO. *See* Food and Agriculture Organization of the United Nations (FAO).

Fassina, Michele. "Aspetti economici e sociali in una grande azienda agricola polesana nel corso del XVII secolo . . ." [Economic and Social Aspects of a Big Polesine Farm in the Course of the Seventeenth Century]. In *Uomini, Terra e Acque, Atti del XIV Convegno di Studi Storici organizzato in collaborazione con l'Accademia del Concordi, Rovigo, 19–20 Novembre 1988* [Men, Land, and Water, Proceedings of the Historical Studies Convention Organized in Collaboration with the Academy of Concordi, Rovigo, Italy, 19–20 November 1988]. Associazione Culturale Minelliana, 1990.

———. "Elementi ed aspetti della presenza del mais nel Vicentino: con particolare riferimento a Lisiera e alla zona attraversata del fiume Tesina" [Elements and Aspects of the Presence of Corn in the Vicentino, with Particular Reference to Lisiera and the Area Traversed by the Tesina River]. In *Lisiera: Immagini, documenti per la storia e cultura di una communità veneta—strutture, congiunture, episodi.* Lisiera, Italy: Edizioni parrocchia de Lisiera, 1981.

———. "L'introduzione della coltura del mais nelle campagne venete" [Introduction of the Cultivation of Corn in the Venetian Countryside]. *Società e Storia* 15 (1982): 31–59.

———. "Il mais nel Veneto nel cinquecento: Testimonianze iconografiche e prime esperienze colturali" [Maize in the Veneto in the Sixteenth Century: Iconographic Testimonies and First Cultural Experience]. In *L'impatto della scoperta dell'America nella cultura Veneziana.* Rome: Bulzone, n.d.

Feeley, James, and Ian Scoones. "Knowledge, Power and Politics: the Environmental Policy-Making Process in Ethiopia." *Journal of Modern African Studies* 38, no. 1 (2000): 89–120.

Fenaroli, Luigi. *Mais, 1946–1967* [Maize, 1946–1967]. Bergamo, Italy: Fondazione Tito Vezio Zapparoli, n.d.

———. *Rapporto sulla sperimentazione maidicola 1953* [Report on Experimentation Maidicola 1953]. Stazione Sperimentale di Maiscoltura, Bergamo, no. 81. Pavia, Italy: Tipografica Ticinese di C. Busca, 1954.

Food and Agriculture Organization of the United Nations (FAO), *Maize in Human Nutrition.* Rome: FAO, 1992.

Franzell, Steven, and Helen van Houten, eds. *Research with Farmers: Lessons from Ethiopia.* Addis Ababa and Wallingford, England: CAB International and Institute of Agricultural Research, 1992.

Friis-Hansen, Esbern. *The Performance of the Seed Sector in Zimbabwe: An Analysis of the Influence of Organizational Structure.* Working Paper no. 66. London: Overseas Development Institute, 1992.

———. *Seeds for African Peasants: Peasants' Needs and Agricultural Research—the Case of Zimbabwe.* Uppsala and Copenhagen: The Nordic Africa Institute and Centre for Development Studies, 1995.

Fussell, Betty. *The Story of Corn: The Myths and History, the Culture and Agriculture, the Art and Science of America's Quintessential Crop.* New York: Knopf, 1992.

Gerhart, John. *The Diffusion of Hybrid Maize in Western Kenya.* Mexico City: CIMMYT, 1975.

Gilbert, Elon. "The Meaning of the Maize Revolution in Sub-Saharan Africa: Seeking Guidance from Past Impacts." Overseas Development Administration Network Paper no. 55, 1995.

Goodman, Major M. "A Brief Survey of the Races of Maize and Current

Attempts to Infer Racial Relationships." In David B. Walden, ed., *Maize Breeding and Genetics*. New York: Wiley and Sons, 1978.

Goodwin, A. J. H. "The Origin of Maize." *South African Archaeological Bulletin* 8 (1953): 13–14.

Gough, David. *A Reconsideration of Some Southern Bantu Maize Terms*. Essays in Bantu Language Studies. Working Paper no. 7. Department of African Languages, Rhodes University, Grahamstown, South Africa, 1981.

Hawke, Sheryl. *Seeds of Change: The Story of Cultural Exchange after 1492*. Menlo Park, Calif.: Addison-Wesley, 1992.

Hay, Margaret Jean. "Economic Change in Luoland: Kowe, 1890–1945," Ph.D. diss., Department of History, University of Wisconsin, Madison, 1972.

Heisey, Paul, and Gregory O. Edmeades. "Maize Production in Drought-Stressed Environments: Technical Options and Research Resource Allocation." In *World Maize Facts and Trends, 1997–1998*. Mexico City: CIMMYT, 1999.

Heisey, Paul, and Wilfred Mwangi. *Fertilizer Use and Maize Production in Sub-Saharan Africa*. Centro Internacional de Mejoridad de Maíz y Trigo (CIMMYT) Economics Working Paper, 96–101. Mexico City: CIMMYT, 1996.

Hoben, Allan. "Family, Land, and Class in Northwest Europe and Northern Highland Ethiopia," *Proceedings of the First United States Conference on Ethiopian Studies*, 1973, 157–170. East Lansing: African Studies Center, Michigan State University, 1975.

ILEIA. *Ecosystem Disruption and Human Health*. Summary report of a consultation hosted by Canada's International Development Research Centre (IDRC) and the United Nations Environment Programme (UNEP). Special supplement to LEISA magazine. Leusden, Netherlands: Centre for Information on Low External Input Sustainable Agriculture, 2002.

Jeffreys, M. D. W. "The History of Maize in Africa." *South African Journal of Science* (March 1954): 197–200.

———. "How Ancient Is West African Maize?" *Africa* 33, no. 2 (1963): 115–131.

———. "Maize Names." *Uganda Journal* 18, 2 (September 1954): 192–193.

———. "The Origin of the Portuguese Word *Zaburro* as Their Name for Maize." *Bulletin de l'Institut Français de l'Afrique Noire*, series B, *Sciences Humaines* 19, nos. 1–2 (January–April 1957): 111–136.

Jennings, Bruce. *Foundations of International Agricultural Research: Science and Politics in Mexican Agriculture.* Boulder, Colo.: Westview, 1988.

Johnston, Bruce. *The Staple Food Economies of Western Tropical Africa.* Stanford, Calif.: Stanford University Press, 1958.

Johnston, Sir Harry. *A Comparative Study of the Bantu and Semi-Bantu Languages.* Oxford: Clarendon Press, 1919–1922.

Jones, W. O. *Manioc in Africa.* Stanford, California: Stanford University Press, 1959.

Juhé-Beaulaton, Dominique. "La diffusion du maïs sur les Côtes de l'Or et des Esclaves aux XVII et XVIII siècles" [The Spread of Maize on the Gold and Slave Coasts in the Seventeenth and Eighteenth Centuries]. *Review Français d'Histoire d'Outre-Mer* 77 (1990): 177–198.

Kassahun Seyoum, Hailu Tafesse, and Steven Franzel. *The Profitability of Coffee and Maize among Smallholders: A Case Study of Limu Awraja, Ilubabor Region.* Research Report no. 10. Addis Ababa: Institute of Agricultural Research, 1990.

Keegan, Timothy. "Crisis and Catharsis in the Development of Capitalism in South African Agriculture." Paper presented to the African Studies Institute, University of the Witwatersrand, Johannesburg, South Africa, October 1984.

Kidane Georgis and Yohannes Degago, eds. *Crop Management Options to Sustain Food Security: Proceedings of the Third Conference of the Agronomy and Crop Physiology Society of Ethiopia, 29–30 May 1997.* Addis Ababa: ACPSE [Agronomy and Crop Physiology Society of Ethiopia], 1998.

Kingdon, Jonathan. *Island Africa: The Evolution of Africa's Rare Animals and Plants.* Princeton, N.J.: Princeton University Press, 1989.

Kitching, Gavin. *Class and Economic Change in Kenya: The Making of an African Petite-Bourgeoisie.* New Haven, Conn.: Yale University Press, 1980,

Knapp, Vincent. "What Europeans Ate in Agricultural Times: Eighteenth-Century Levels of Food Consumption and Nutrition," 2–6. Paper presented to the Seminar in Agrarian Studies, Yale University, 22 January, 1999.

Koelle, Sigismund Wilhelm. *Polyglotta Africana,* P. E. H. Hair and David Dalby, eds. Graz, Austria: Akademische Druk-U., 1963.

Kokwe, Gun Mickels. *Maize, Markets and Livelihoods: State Intervention and Agrarian Change in Luapula Province, Zambia, 1950–1995.* Helsinki: Interkont Books, 1998.

Langlands, B. W. "Maize in Uganda." *Uganda Journal* 29, no. 2 (1965): 215–221.

Levi, G. "Innovazione tecnica e resistenza contadina: Il mais nel Piemonte del '600" [Technical Innovation and Peasant Resistance: Maize in the Piedmont in the Seventeenth Century]. *Quaderni Storici* 42 (September–December 1979): 1092–1100.

Lipton, Michael, with Longhurst Richard. *New Seeds and Poor People.* London: Unwin Hyman, 1989.

Mangelsdorf, Paul C. *Corn: Its Origin, Evolution, and Improvement.* Cambridge, Mass.: Harvard University Press, 1974.

Masefield, G. B. "Maize Names: Correspondence." *Uganda Journal* 14, no. 1 (March 1950): 106.

Mashingaidze, Kingston. "Maize Research and Development." In Mandivamba Rukuni and Carl Eicher, eds., *Zimbabwe's Agricultural Revolution.* Harare: University of Zimbabwe Press, 1997.

Masters, William A. *Government and Agriculture in Zimbabwe.* Westport, Conn.: Greenwood, 1994.

McCann, James C. *Green Land, Brown Land, Black Land: An Environmental History of Africa* (Portsmouth, N.H., and Oxford: Heinemann and James Currey, 1999).

———. "Maize and Grace: History, Corn, and Africa's New Landscapes, 1500–1999." *Comparative Studies in Society and History* 43, no. 2 (April 2001): 246–272.

———. *People of the Plow: An Agricultural History of Ethiopia.* Madison: University of Wisconsin Press, 1995.

McClintock, Barbara. "Significance of Chromosome Constitutions in Tracing the Origin and Migration of Races of Maize in the Americas." In David B. Walden, ed., *Maize Breeding and Genetics.* New York: Wiley and Sons, 1978.

McIntire, John, Danile Bourzat, and Prabhu Pingali. *Crop-Livestock Interactions in Sub-Saharan Africa.* Washington, D.C.: World Bank, 1990.

McNeill, William H. "American Food Crops in the Old World." In Herman J. Viola and Carolyn Margolis, eds., *Seeds of Change: A Quincentennial Commemorative.* Washington, D.C.: Smithsonian Institution Press, 1991.

Messedaglia, Luigi. *Il mais e la vita rurale Italiana: Saggio di storia agraria.* Piacenza, Italy: Federazione Italiana dei Consorzi Agrari, 1927.

Mhike, Xavier. "The History of Maize Breeding in Zimbabwe: Methods and Tools." SIRDC [Scientific and Industrial Research and Development Center, Zimbabwe] Biotechnology Proceedings, May 2003.

Miller, David, and Wally Marasas. "Ecology of Mycotoxins in Maize and Groundnuts." In ILEIA, *Ecosystem Disruption and Human Health*. Leusden, Netherlands: Centre for Information on Low External Input Sustainable Agriculture, 2002.

Mintz, Sidney W. *Sweetness and Power*. New York: Viking, 1985.

Miracle, Marvin P. "Interpretation of Evidence on the Introduction of Maize into West Africa." *Africa* 33, no. 2 (1963): 2–35.

———. *Maize in Tropical Africa*. Madison: University of Wisconsin Press, 1966.

Montelli, Roberto. *Le piante erbacee del nuovo mondo nella storia dell'agricoltura Italiana* [Herbaceous Plants of the New World in the History of Italian Agriculture]. Serie di Storia Economica, no. 1. Genoa: Istituto di Studi Economici, Università degli Studi di Genova, 1994.

Morris, Michael, ed. *Maize Seed Industries in Developing Countries*. Boulder, Colo., and Mexico City: Lynne Rienner and CIMMYT, 1998.

Munro, William. "To Civilize Both the Land and the People: Governmentality and the Environment in Late-Colonial Zimbabwe." Paper presented to the annual meeting of the American Society for Environmental History, Providence, R.I., March 2003.

Muratori, Carlo. "Maize Names and History: A Further Discussion." *Uganda Journal* 16, no. 1 (March 1952): 76–81.

Murray, Colin. *Black Mountain: Land, Class, and Power in the Eastern Orange Free State, 1880s to 1980s*. Washington, D.C.: Smithsonian Institution Press, 1992.

———. *Families Divided: The Impact of Labour in Lesotho History*. Johannesburg: Raven Press, 1981.

Myers, O., Jr. "Breeding for Drought Tolerance in Maize." In Brhane Gelaw, ed., *To Feed Ourselves: A Proceedings of the First Eastern, Central, and Southern African Regional Workshop*, Lusaka, Zambia, March 1985. Mexico City: CIMMYT, 1985.

National Research Council. *Postharvest Food Losses in Developing Countries*. Washington, D.C.: National Academy of Sciences, 1978.

Natrass, R. M. "Occurrence of *Puccinia polysora Underw.* in East Africa," *Nature* 171 (1952): 527.

Ngwira, L. D. M., and E. M. Sibale. "Maize Research and Production in Malawi." In Brhane Gelaw, ed., *To Feed Ourselves: A Proceedings of the First Eastern, Central, and Southern African Regional Workshop*, Lusaka, Zambia, March 1985. Mexico City: CIMMYT, 1985.

Noacco, C. "Aspetti nutrizionali della polenta." In Camera di Commercio, Industria, Artigianato e Agricoltura di Udine, *Polenta di qualità in Friuli*.

Udine, Italy: CCIAA [Camera di Commercio, Industria, Artigianato e Agricoltura], 1987.

Nweke, Felix I., Dunstan Spencer, and John Lynam. *The Cassava Transformation: Africa's Best-Kept Secret.* East Lansing: Michigan State University Press, 2002.

Olver, R. C. "The Zimbabwe Maize Breeding Program." In Brhane Gelaw, ed., *To Feed Ourselves: A Proceedings of the First Eastern, Central, and Southern African Regional Workshop,* Lusaka, Zambia, March 1985. Mexico City: CIMMYT, 1985.

————. "Zimbabwe Maize Breeding Program." In Brhane Gelaw, ed., *Towards Self-Sufficiency: A Proceedings of the Second Eastern, Central and Southern Africa Workshop.* Mexico City: CIMMYT, 1988.

Paarlberg, Robert. "Africa and the Global GM Crop Conflict: Why Europe Will Win, and Poor Farmers Lose." Working Paper no. 245, African Studies Center, Boston University, 2003.

Pasch, Helma. "Zur Geschichte der Verbreitung des Maises in Africa" [On the History of the Spread of Maize in Africa]. *Sprache und Geschichte in Afrika 5* (1983): 177–218.

Patrizia, Paolo. *Terra patrizia: Aristocrazie terriere e società rurale in Veneto e Friuli* [Patrician Earth: The Landed Aristocracy and Rural Society in the Veneto and Friuli]. Venice: Istituto Editoriale Veneto-Friulano, 1993.

Pellegrini, Giovan Battista, ed. *I lavori dei contadini* [The Work of the Peasants]. Vicenza, Italy: Neri Pozza, 1997.

Pelligrini, Giovan Battista, and Carla Marcato. "Granoturco." In G. B. Pellegrini and C. Marcato, *Terminologia Agricola Friulana.* Udine, Italy: Società Filologica Friuliana, 1992.

Perkins, John, H. *Geopolitics and the Green Revolution: Wheat, Genes, and the Cold War.* Oxford: Oxford University Press, 1997.

Peters, Pauline. "The Limits of Knowledge: Securing Rural Livelihoods in a Situation of Resource Scarcity." In C. B. Barrett, F. Place, and A. A. Aboud, eds., *Natural Resource Management Practices in African Agriculture: Understanding and Improving Current Practices.* Cambridge, Mass.: CABI Publishing, 2002.

Phillips, John, John Hammond, L. H. Samuels, and R. J. M. Swynnerton. *The Development of the Economic Resources of Southern Rhodesia with Particular Reference to the Role of African Agriculture.* Report of the Advisory Committee. Salisbury, Southern Rhodesia, 1962.

Pingali, P. L., ed. *CIMMYT 1999–2000 World Maize Facts and Trends. Meeting World Maize Needs: Technological Opportunities and Priorities for the Public Sector.* Mexico City: CIMMYT, 2001.

Portères, Roland. "L'introduction du maïs en Afrique." *Journal de l'Agriculture Tropicale et de Botanique Appliquée* 2, nos. 5–6 (May-June 1955): 221–231.

Ransom, J. K., A. F. E. Palmer, B. T. Zambezi, Z. O. Mduruma, S. R. Waddington, K. V. Pixley, and D. C. Jewell. *Maize Productivity Gains through Research and Technology Dissemination: Proceedings of the Fifth Eastern and Southern Africa Regional Maize Conference*, 3–7 June 1996, Arusha. Addis Ababa: CIMMYT, 1997.

Rhind, D., J. M. Waterston, and F. C. Deighton. "Occurrence of *Puccinia polysora* Underw. in West Africa." *Nature* 169 (1952): 631.

Richards, Audrey I. *Land, Labour, and Diet in Northern Rhodesia: An Economic Study of the Bemba Tribe*. London: Oxford University Press, 1939.

Ristanovic, D., and P. Gibson. "Development and Evaluation of Maize Hybrids in Zambia." In Brhane Gelaw, ed., *To Feed Ourselves: A Proceedings of the First Eastern, Central, and Southern African Regional Workshop*, Lusaka, Zambia, March 1985. Mexico City: CIMMYT, 1985.

Ristanovic, D., O. Meyers, W. Mwale, C. Mwambula, and P. Magnuson. "Maize Inbred Line Development and the Role of Population Improvement in Zambia." In Brhane Gelaw, ed., *Towards Self-Sufficiency: A Proceedings of the Second Eastern, Central and Southern Africa Workshop*. Mexico City: CIMMYT, 1988.

Rizzolatti, Piera. "La stalla e il governo degli animali" [The Stall and the Management of Animals]. In Giovan Battista Pellegrini, ed., *I lavori dei contadini*. Vicenza, Italy: Neri Pozza, 1997.

Roberts, Bruce D. "The Incorporation of Maize into Africa." In Leonard Plotnikov and Richard Scaglion, eds., *The Globalization of Food*. Prospect Heights, Ill.: Waveland Press, 1999.

Rosetti, V. "Breve storia del mais in Friuli" [A Short History of Maize in Friuli]. In Camera di Commercio, Industria, Artigianato e Agricoltura di Udine, *Polenta di qualità in Friuli*. Udine, Italy: CCIAA [Camera di Commercio, Industria, Artigianato e Agricoltura], 1987.

Rouanet, Guy. *Maize*. Tropical Agriculturalist Series. London: Macmillan, 1995.

Rukuni, Mandivamba, and Carl Eicher, eds. *Zimbabwe's Agricultural Revolution*. Harare: University of Zimbabwe, 1997.

Sasakawa Africa Association. *Sasakawa Global 2000 Agricultural Project in Ethiopia*. Mexico City: Sasakawa Africa Association, 1995.

Scarpa, Giorgio. *Il mais nell'economia agraria Italiana*. [Maize in Italian Agrarian Economy]. Stazione Sperimentale di Maiscoltura, Bergamo, no. 79. Venice: Fantoni, 1954.

Schweinfurth, G., ed. *Emin Pasha in Central Africa*. London: Phillip and Son, 1888.

Secretary of Agriculture, Southern Rhodesia. *Report of the Secretary of Agriculture (Southern Rhodesia) for the Year of 1962*. Salisbury, Southern Rhodesia: Ministry of Agriculture, 1962.

Singh, R. P., Suresh Pal, and M. Morris. *Maize Research and Development and Seed Production in India: Contributions of the Public and Private Sectors*. CIMMYT Economics Working Paper no. 95–03. Mexico City: CIMMYT, 1995.

Smale, Melinda. "'Maize Is Life': Malawi's Delayed Green Revolution." *World Development* 23, no. 5 (1995): 819–831.

Smale, Melinda, and Paul Heisey. "Maize Research in Malawi: An Emerging Success Story?" *Journal of International Development* 6, no. 6 (1994): 689–706.

Smale, Melinda, Paul Heisey, and Howard D. Leathers. "Maize of the Ancestors and Modern Varieties: The Microeconomics of High-Yielding Variety Adoption in Malawi." *Economic Development and Cultural Change* 43, no. 2 (1995): 351–368.

Smale, Melinda, and Thomas Jayne. *Maize in Eastern and Southern Africa: "Seeds" of Success in Retrospect*. EPTD Discussion Paper no. 97. Washington, D.C.: Environmental and Technical Division, International Food Policy and Research Institute, January 2003.

Smith, J. J., Georg Weber, M. V. Manyong, and M. A. B. Fakorede. "Fostering Sustainable Increases in Maize Productivity in Nigeria." In Derek Byerlee and Carl Eicher, eds., *Africa's Emerging Maize Revolution*. Boulder, Colo.: Lynne Rienner, 1997.

Smith, R. Cherer. *The Story of Maize and the Farmers' Co-op Ltd*. Salisbury, Southern Rhodesia: Farmers' Co-op, 1979.

Sorenson, John L., and Martin H. Raish. *Pre-Columbian Contact with the Americas across the Oceans: An Annotated Bibliography*. 2 vols. Provo, Utah: Research Press, 1990.

Stanton, W. R., and R. H. Cammack, "Resistance to the Maize Rust, *Puccinia polysora Underw*." *Nature* 172 (1953): 505–506.

Storey, H. H., and A. K. Howland. "Resistance in Maize to the Tropical American Rust Fungus *Puccinia polysora Underw*." *Heredity* 11 (1957): 289–301.

Taba, S., ed. *Maize Genetic Resources*. Maize Program Special Report. Mexico City: CIMMYT 1995.

Tattersfield, Rex J., and Ephraim K. Havazvidi, "The Development of the Seed Industry." In Mandivamba Rukuni and Carl Eicher, eds., *Zimba-*

bwe's *Agricultural Revolution*. Harare: University of Zimbabwe Press, 1997.

Tawonezvi, Patrick. "Agricultural Research Policy." In Mandivamba Rukuni and Carl Eicher, eds., *Zimbabwe's Agricultural Revolution*. Harare: University of Zimbabwe, 1997.

Tesfaye Tsegaye and Alemu Hailye. *Adoption of Improved Maize Technologies and Inorganic Fertilizer in Northwestern Ethiopia*. Research Report no. 40. Addis Ababa: Institute for Agricultural Research, 2001.

Tomini, D. "Mais nel mondo: Alcuni piatti-tradizione" [Maize in the World: Some Traditional Dishes]. In Camera di Commercio, Industria, Artigianato e Agricoltura di Udine, *Polenta di qualità in Friuli*. Udine, Italy: CIAA [Camera di Commercio, Industria, Artigianato e Agricoltura], 1987.

Tripp, Robert. *Seed Provision and Agricultural Development: The Institutions of Rural Change*. London, Oxford, and Portsmouth, NH: Overseas Development Institute, James Currey, and Heinemann, 2001.

Tripp, Robert, and Kofi Marfo. "Maize Technology Development in Ghana during Economic Decline and Recovery." In Derek Byerlee and Carl Eicher, eds., *Africa's Emerging Maize Revolution*. Boulder, Colo.: Lynne Rienner, 1997.

Van Eijnatten, C. L. M. *Towards the Improvement of Maize in Nigeria*. Wageningen, Netherlands: Veenman and Zonen N. V., 1965.

Vickery, Kenneth. *Black and White in Southern Zambia: The Tonga Plateau Economy and British Imperialism, 1890–1939*. Westport, Conn.: Greenwood Press, 1986.

———. "Saving Settlers: Maize Control in Northern Rhodesia." *Journal of Southern African Studies* 11, no. 2 (April 1985): 212–234.

Viola, Herman J., and Carolyn Margolis. *Seeds of Change: A Quincentennial Commemorative*. Washington, D.C.: Smithsonian Institution Press, 1991.

Warman, Arturo. *La historia de un bastardo: Maíz y capitalism* [The History of a Bastard: Maize and Capitalism]. Mexico City: Fondo de Cultura Económica, 1988.

Watt, Ian. "Regional Patterns of Cereal Production and Consumption." In Zein Ahmed Zein and Helmut Kloos, eds., *The Ecology of Health and Disease in Ethiopia*. Addis Ababa: Ministry of Health, 1988.

Wedderburn, R., K. Short, and H. Pham. "Progress in Maize Improvement at CIMMYT's Zimbabwe Maize Research Station." In Brhane Gebre Kidan, ed., *Maize Improvement, Production, and Protection in Eastern and Southern Africa: Proceedings of the Third Eastern and Southern African Regional Maize Workshop*. Mexico City: CIMMYT, 1990.

Wellhausen, E. J. "Recent Developments in Maize Breeding in the Tropics." In David B. Walden, ed., *Maize Breeding and Genetics*. New York: Wiley and Sons, 1978.

Willett, Frank. "The Introduction of Maize into West Africa: An Assessment of Recent Evidence." *Africa* 32, no. 1 (1962): 1–13.

Woldeyesus Sinebo, ed. *Crop Management Research for Sustainable Production: Status and Potentials. Proceedings of the Second Annual Conference of the Agronomy and Crop Physiology Society of Ethiopia*. Addis Ababa: ACPSE [Agronomy and Crop Physiology Society of Ethiopia], 1997.

Woldeyesus Sinebo, Zerihun Tadele, and Nigussie Alemayehu. *Increasing Food Production through Improved Crop Management: Proceedings of the First and Inaugural Conference of the Agronomy and Crop Physiology Society of Ethiopia*. Addis Ababa: ACPSE [Agronomy and Crop Physiology Society of Ethiopia], 1996.

Wright, A. C. A. "Maize Names as Indicators of Economic Contacts." *Uganda Journal* 13, no. 1 (March 1949): 61–81.

Yemane Ye-ebiyo, Richard Pollack, Anthony Kiszewski, and Andrew Spielman. "A Component of Maize Pollen That Stimulates Larval Mosquitoes *(Diptera culicidae)* to Feed and Increases Toxicity of Microbial Larvicides." *Journal of Medical Entomology* 40, no. 6 (November 2003): 860–864.

———. "Enhancement of Development of Larval *Anopheles arabiensis* by Proximity to Flowering Maize *(Zea mays)* in Turbid Water and When Crowded." *American Journal of Tropical Medicine and Hygiene* 68, no. 6 (2003): 748–752.

———. "Enhanced Development in Nature of Larval *Anopheles arabiensis* Mosquitos Feeding on Maize Pollen." *American Journal of Tropical Medicine and Hygiene* 63, nos. 1–2 (2000): 90–93.

Yeshanew Ashagrie, Matts Olsen, and Tekalign Mamo. "Contribution of *Croton macrostachys* to Soil Fertility in Maize-Based Subsistence Agriculture of Bure [Burie] Area, North-Western Ethiopia," *Maize Production Technology for the Future: Challenges and Opportunities: Proceedings of the Sixth Eastern and Southern Africa Regional Maize Conference*. Addis Ababa: CIMMYT, 1999.

Zalin, Giovanni. *La società agraria Veneta del secondo ottocento: Possidenti e contadini nel sottosviluppo regionale* [The Agrarian Society of the Veneto in the Eighteenth Century: Landowners and Peasants in Regional Development]. Padua, Italy: Casa Editrice Dott. Antonio Milani, 1978.

Acknowledgments

People cling to an appealing image of the historian's craft as a solitary one, of the individual scholar laboring in the darkened recesses of a study, a library, or an archive. The task of that apocryphal scholar is to weave a story from a welter of facts and ideas. The reality of reconstructing history at the outset of the new millennium is, however, quite different. The historian's task is the telling of a story that explains why and how, but the process is one that involves, in addition to inspiration, a great many people, institutions, and sources of emotional and financial support. Such is the case with *Maize and Grace*. This book had its genesis in farmers' fields in Ethiopia, in research laboratories, around grain silos, in seminars, in libraries, and in dusty archives. Many people from all those settings—certainly even more than the few that I can mention here—deserve heartfelt thanks and acknowledgment.

Funding is critical to any research enterprise, and this one is no exception. This project has received generous support from Fulbright-Hays, for my research in Britain, Italy, Ethiopia, and Zimbabwe; from the Ford Foundation, for research at the Centro Internacional de Mejoridad de Maíz y Trigo (CIMMYT, International Maize and Wheat Improvement Center) in Mexico City, at the Public Records Office in London, and at the Royal Botanic Gardens, Kew; from the Boston University International Program in Padua, Italy; from the Rockefeller Foundation, for visits to Tanzania and Lesotho; and from Yale University, for a productive year of research in its Program in Agrarian Studies.

Financial resources are important, but people provide essential hospitality, inspiration, moral support, and kindness in varying measures. At Boston University, my institutional home since 1984, the people at the African Studies Center have inspired and supported me; those folks include John Harris, Joanne Hart, Jean Hay, James Anthony Pritchett, Parker Shipton, and Diana Wylie. In the Department of History I owe many debts to colleagues, some of whom know about maize and many who do not but have inspired me nonetheless. That list includes Charles Dellheim, Barbara Diefendorf, Saul Engelbourg, Louis Ferleger, Tom Glick, and Sarah Phillips. My botanist colleague Gillian Cooper-Driver helped with introductions at her former place of employment Kew Gardens in London. Deans and provosts also offered special support, especially Provost Dennis Berkey and Dean Jeffrey Henderson. Boston is such a stupendous place for research because of the geographic concentration of wonderful colleagues. At Harvard Pauline Peters and Andy Spielman offered advice and ideas. At MIT once a month, Deborah Fitzgerald and the participants in the Modern Times, Rural Places seminar offered me a home across the Charles.

Elsewhere in the United States I enjoyed support and ideas from Bob Harms, Cassandra Moseley, and Kay Mansfield at the Yale Program in Agrarian Studies. Christopher Ehret, a unique font of linguistic knowledge, was generous with his time. In East Lansing, Thom Jayne offered insights from his own pathbreaking approaches to agricultural economics.

In Ethiopia I owe great debts to new friends and long-standing ones. At Addis Ababa University Baye Yimam, Gebre Selassie, Shiferaw Bekele, Teferi Abate, Tekalign Wolde Mariam, and Tekle Haimanot were most helpful. At the CIMMYT offices I learned constantly from Douglas Tanner, S. Twumasi-Afriye, and Benti Tolossa and received key introductions from Tom Payne. For my initiation into the science of malaria and the impact of the disease on human beings, I owe special thanks to Asnakew Kebbede, Yemane Ye-ebiyo, Tesfaye Zewde, and my old friends Semahagne Abate and Haile Semahagne. At Sasakawa Global 2000, Tekele Gebru taught me a great deal and also listened.

In Zimbabwe I was generously hosted by Dr. Marianne Bänziger at CIMMYT-Harare. This innovative maize breeder also introduced me to Cosmos Magorokosho and his work on local maize biodiversity. Mike Caulfield generously shared his memories and documents of early fieldwork in Rhodesia. Elizabeth Nugent was a gracious host in Harare.

In Ghana my brief visit in 1996 was especially productive because of the efforts of my historian colleague Kofi Baku at Legon and of Ben Dzah at the Crop Research Institute in Kumasi. At CIMMYT headquarters at El Batán, Mexico, I received wise counsel from Norman Borlaug, Christopher Dowswell, and Mauricio Bellon.

In Italy I visited several important research sites and profited from the helpful advice of a number of scholars of *mais italiano,* including my friend Armando DeGuio at the Università de Padua, Mauro Tosca at the Università di studi di Napoli (Orientale), as well as Marco Bertolini and Alberto Vederio at the Istituto per la Cerealicoltura in Bergamo. My old friend Alemneh Dejene arranged introductions for me at the Food and Agriculture Organization of the United Nations (FAO) in Rome.

I made several trips to Britain to work at the Public Records Office and Kew Gardens. On each of those London visits I benefited from the hospitality of my musician friends Patsy and John Kitchen. In my digs on the South Downs in Sussex I was often the recipient of sage advice and familial hospitality from James Fairhead and Melissa Leach in Rotten Row in Lewes and Poverty Bottom in Norton and enjoyed wonderful accommodations organized by Jean Hudson in Kingston. David Anderson and Richard Grove were valuable colleagues there and in London. At Kew Gardens, Dr. Brian Spooner was instrumental in my gaining access to the Kew library and the facilities of the Mycology Laboratory. Even though Joyce Seltzer and Susan Abel, my editors at Harvard University Press, joined this effort late in the process, they were admirable guides in framing the story.

Last of all, those who know me are aware of the enormous debt that I owe to my family. Martha and Libby, my daughters, are almost ready to set forth on their own distinctive life paths. My wife, Sandi, is perhaps the most valuable of all of my colleagues and supporters. Her love and dedication to our common tasks in life has been most important to me of all.

Illustration Credits

Index